Complex Systems Design & Management

Gérard Auvray · Jean-Claude Bocquet
Eric Bonjour · Daniel Krob
Editors

Complex Systems Design & Management

Proceedings of the Sixth International Conference on Complex Systems Design & Management, CSD&M 2015

Springer

Editors
Gérard Auvray
Airbus Defense and Space
Les Mureaux
France

Eric Bonjour
Université de Lorraine
Nancy
France

Jean-Claude Bocquet
ECP Grande Voie Des Vignes
Chatenay-Malabry
France

Daniel Krob
CESAMES
Paris
France

ISBN 978-3-319-26107-2 ISBN 978-3-319-26109-6 (eBook)
DOI 10.1007/978-3-319-26109-6

Library of Congress Control Number: 2015953795

Springer Cham Heidelberg New York Dordrecht London

Printed on acid-free paper

Springer International Publishing AG Switzerland is part of Springer Science+Business Media
(www.springer.com)

Preface

Introduction

This volume contains the proceedings of the Sixth International Conference on "Complex Systems Design & Management Paris" (CSD&M Paris 2015; see the conference Web-site: http://www.csdm2015.csdm.fr for more details).

The CSD&M Paris 2015 conference was organized on November 23–25, 2015, at the EDF Lab at Clamart (France) under the High Patronage of the French Ministry of Economy, Industry and Digital. The two following founding partners are given below:

1. The non-profit organization C.E.S.A.M.E.S. (Center of Excellence on Systems Architecture, Management, Economy and Strategy),
2. The Ecole Polytechnique—ENSTA ParisTech—Télécom ParisTech—Dassault Aviation—DCNS—DGA—Thales "Engineering of Complex Systems" chair.

The conference benefited of the permanent support of many academic organizations such as Ecole Polytechnique, CentraleSupélec, ENSTA ParisTech and Télécom ParisTech which were deeply involved in its organization.

We also would like to thank the conference partners: Airbus Group, CEA LIST, Dassault Aviation, Dassault Systemes, DCNS, Digiteo Labs, Direction Générale de l'Armement (DGA), Faurecia, l'IRT SystemX, MEGA International, Obeo, PPI, and Thales which were the main industrial and institutional sponsors of the conference. The generous specific support of Airbus Group and Dassault Systemes shall be especially pointed out here.

We are also grateful to several non-profit organizations such as Association Française d'Ingénierie Système (AFIS), International Council on Systems Engineering (INCOSE) and Systematic Paris Region Systems & ICT Cluster which strongly supported our communication effort.

Finally, our special thanks go to EDF which offers us to welcome the CSD&M Paris 2015 conference at EDF Lab, in Clamart.

Why a CSD&M Conference?

Mastering complex systems requires an integrated understanding of industrial practices as well as sophisticated theoretical techniques and tools. This explains the creation of an annual *go-between* forum at European level (which did not existed yet) dedicated both to academic researchers and to industrial actors working on complex industrial systems architecture and engineering. Facilitating their *meeting* was actually for us a *sine qua non* condition in order to nurture and develop in Europe the science of systems which is currently emerging.

The purpose of the "Complex Systems Design & Management" Paris (CSD&M Paris) conference is exactly to be such a forum, in order to become, in time, *the* European academic-industrial conference of reference in the field of complex industrial systems architecture and engineering, which is a quite ambitious objective. The last five CSD&M Paris 2010, CSD&M Paris 2011, CSD&M Paris 2012, CSD&M Paris 2013, and CSD&M Paris 2014 conferences—which were held in the end of October 2010, December 2011, December 2012, December 2013, and November 2014 in Paris—were the first steps in this direction. In 2014, there were almost 300 participants who came from 20 different countries, which measures the growing success of the CSD&M Paris conference.

Our Core Academic—Industrial Dimension

To make the CSD&M Paris conference this convergence point of the academic and industrial communities in complex industrial systems, we based our organization on a principle of *complete parity* between academics and industrialists (see the conference organization sections in the next pages). This principle was first implemented as follows:

- the Program Committee consisted of 50 % academics and 50 % industrialists,
- the Invited Speakers came in a balanced way from numerous professional environments.

The set of activities of the conference followed the same principle. They indeed consist of a mix of research seminars and experience sharing, academic articles and industrial presentations, software and training offers presentations, etc. The conference topics cover in the same way the most recent trends in the emerging field of complex systems sciences and practices from an industrial and academic perspective, including the main industrial domains (aeronautic and aerospace, transportation and systems, defense and security, electronics and robotics, energy and environment, healthcare and welfare services, media and communications, software and e-services), scientific and technical topics (systems fundamentals, systems architecture and engineering, systems metrics and quality, systemic tools), and system types (transportation systems, embedded systems, software and information systems, systems of systems, artificial ecosystems).

The 2015 Edition

The CSD&M Paris 2015 Edition received 75 submitted papers, out of which the Program Committee selected 20 regular papers to be published in the conference proceedings. A 27 % acceptance ratio has been reached and guarantees the high quality of the presentations. The Program Committee also selected 20 papers for a collective presentation during the poster workshop of the conference.

Each submission was assigned to at least two Program Committee members, who carefully reviewed the papers, in many cases with the help of external referees. These reviews were discussed by the Program Committee during a physical meeting held in C.E.S.A.M.E.S. office in Paris by the May 29, 2015, and via the EasyChair conference management system.

We also chose 12 outstanding speakers with various industrial and scientific expertises who gave a series of invited talks covering all the spectrum of the conference, mainly during the two first days of CSD&M Paris 2015. The first and second days of the conference were especially organized around a common topic—Smart Cities: Systems Issues—that provided consistency to all invited talks. The last day was finally dedicated to a special "thematic session," followed by a "à la carte" program: the presentations of the 20 accepted papers, the conference partners' workshops, an IBM workshop, and an IEEE Chapter workshop. Furthermore, we had a poster workshop for encouraging presentations and discussions on interesting but "not-yet-polished" ideas. CSDM Paris 2015 also offered a Systems Architecture & Engineering Tools Session in order to provide each participant a good vision of the last engineering and technological news.

Paris
November 2015

Gérard Auvray
Jean-Claude Bocquet
Eric Bonjour
Daniel Krob

Conference Organization

Conference Chairs

- **General Chair**
  - Daniel Krob, Incose Fellow, C.E.S.A.M.E.S. & Ecole Polytechnique, France
- **Organizing Committee Chair**
  - Jean-Claude Bocquet, CentraleSupélec, France
- **Program Committee Chairs**
  - Gérard Auvray, Airbus Defense and Space, France
  - Eric Bonjour, Ensgsi Université De Lorraine, France

Program Committee

The Program Committee consists of 32 members (17 academic and 15 industrial) of high international visibility. Their spectrum of expertise covers all of the conference topics.

Academic Members

- *Co-chair*
  - Eric Bonjour, Ensgsi Université De Lorraine, France
- *Other Members*
  - Vincent Chapurlat, Ecole des Mines d'Alès, France
  - Eric Coatanea, Aalto University, Finland
  - Olivier De Weck, MIT, USA
  - Timothy Ferris, Unisa, Australia

- Alfredo Garro, University of Calabria, Italy
- Nil Handen Ergin, Penn State University, USA
- Michael Henshaw, Loughborough University Leicestershire, Great Britain
- Paulien Herder, University of Delft, Netherlands
- Claude Y. Laporte, École de Technologie Supérieure, Canada
- Jörg Lalk, University of Pretoria, South Africa
- Eric Levrat, Université de Lorraine/Cran, France
- Maik Mauer, Technische Universität München, Germany
- Patrick Millot, Université Valenciennes, France
- Chris Paredis, Georgia Institute of Technology, USA
- Brian Sauser, University of North Texas, USA
- Dinesh Verma, Stevens Institute of Technology, USA

Industrial Members

- *Co-chair*
  - Gérard Auvray, Airbus Defense and Space, France
- *Other Members*
  - Henrik Balslev, Systems Engineering Denmark Aps, Denmark
  - Richard Beasley, Rolls Royce, United Kingdom
  - Jean-Pierre Daniel, Areva, France
  - Alain Dauron, Renault, France
  - Gauthier Fanmuy, Dassault Systemes, France
  - Sanford Friedenthal, Lockheed-Martin, USA
  - Mike Johnson, Ruag, Switzerland
  - Juan Llorens, The Reuse Company, Spain
  - David Long, Vitech, Canada
  - Clotilde Marchal, Airbus Group, France
  - Garry Roedler, Lockheed-Martin, USA
  - Jean-Claude Roussel, Airbus Group Innovation, France
  - Sven-Olaf Schulze, LINITY AG, Germany
  - Robert Swarz, MITRE, USA

Organizing Committee

The Organizing Committee consists of 18 members (academic and industrial) of high international visibility. The Organizing Committee is in charge of defining the agenda/program of the conference, identifying keynotes speakers and has to ensure the functioning of the event (sponsoring, communication…). Its spectrum of expertise covers all of the conference topics.

Organizing Committee

- *Chair*
  - Jean-Claude Bocquet, CentraleSupélec, France
- *Other Members*
  - Marc Aiguier, Ecole Centrale de Paris, France
  - Anas Alfaris, CCES & MIT, Saudi Arabia
  - Emmanuel Arbaretier, APSYS, Airbus Group, France
  - Frédéric Boulanger, CentraleSupélec, France
  - Guy Boy, Florida Institute of Technology, USA
  - Cihan Dagli, Missouri University of Science and Technology, USA
  - John Fitzgerald, Newcastle University, United Kingdom
  - Pascal Foix, Thales, France
  - Eric Goubault, CEA LIST, France
  - Chahinez Hamlaoui, Systematic Paris Region Systems & ICT Cluster, France
  - Paul Labrogere, IRT SystemX, France
  - Frédéric Magoules, Ecole Centrale Paris, France
  - Garry Roedler, Lockheed Martin Corporate Engineering, USA
  - François Stephan, IRT SystemX, France
  - Nicolas Trêves, CNAM, France
  - Jon Wade, Stevens Institute of Technology, USA
  - David Walden, Sysnovation & INCOSE, USA

Invited Speakers

Grand Challenges—Society

- Serge Salat, President, Urban Morphology and Complex Systems Institute, Paris, France
- Daniel Kaplan, President, The Next-Generation Internet Foundation, Europe
- Carlos Moreno, COFELY INEO, Professor, President Scientific Advisor, FSIM COFELY Department (INEO, AXIMA, ENDEL) the Energy Services business line, GDF Suez, France

Grand Challenges—Industry

- Pascal Terrien, Director, Sustainable Cities Program, EDF R&D, France & Singapore
- Pierre Guehenneux, Digital Transformation Director, Vinci Construction, France

Scientific State of the Art

- Eric Goubault, Institute Professor and Director of the "Complex Systems Engineering" Chair, Ecole Polytechnique, France
- Joseph Sifakis, Professor, Ecole Polytechnique Fédérale de Lausanne, Switzerland
- Sébastien Tremblay, Scientific Director of UMR in Urban Scholars of Laval University, Canada

Methodological State of the Art

- Laurent Schmitt, Vice-President Innovation and Strategy, Alstom Grid, France
- Michel-Alexandre Cardin, Professor, National University of Singapore, Singapore
- Jakob Puchinger, Head of the business unit dynamic transportation systems, Austrian Institute of Technology, Austria

Acknowledgments

We would like to thank all members of the Program and Organizing Committees for their time, effort, and contributions to make CSD&M Paris 2015 a top-quality conference. A special thank is addressed to the C.E.S.A.M.E.S. non-profit organization team which managed permanently with an huge efficiency all the administration, logistics, and communication of the CSD&M Paris 2015 conference (see http://www.cesames.net).

The organizers of the conference are also greatly grateful to the following sponsors and partners without whom the CSD&M Paris 2015 event would not exist:

- **Founding Partners**
  - C.E.S.A.M.E.S.—Center of Excellence on Systems Architecture, Management, Economy and Strategy,
  - Ecole Polytechnique—ENSTA ParisTech—Télécom ParisTech—Dassault Aviation—DCNS—DGA—Thales "Engineering of Complex Systems" Chair.
- **Academic Sponsors**
  - CEA LIST,
  - CNAM—Conservatoire National des Arts et Métiers,
  - Ecole Polytechnique,
  - CentraleSupélec,
  - ENSTA ParisTech,
  - Télécom ParisTech.
- **Industrial and Institutional Sponsors**
  - Airbus Group,
  - Dassault Aviation,

- Dassault Systemes,
- DCNS,
- Direction Générale de l'Armement (DGA),
- EDF,
- Faurecia,
- MEGA International,
- Obeo,
- PPI,
- Thales.

- **Institutional Sponsors**
  - Digiteo labs,
  - IRT SystemX,
  - French Ministry of Economy, Industry and Digital.
- **Supporting Partners**
  - Association Française d'Ingénierie Système (AFIS),
  - International Council on Systems Engineering (INCOSE),
  - Systematic Paris Region Systems & ICT Cluster.
- **Participating Partners**
  - Ellidiss,
  - Esterel Technologies,
  - IBM Analytics,
  - IEEE Chapter,
  - Knowledge Inside,
  - No Magic Europe MBSE,
  - Pragmadev,
  - PTC,
  - Squoring Technologies,
  - The CoSMo Company,
  - The MathWorks.

At the publishing moment, new partners could join CSDM Paris 2015; you can find them on the conference Web-site.

Contents

Part II Posters

Part I
Regular Papers

Lessons Learnt in System Engineering for the SESAR Programme

Alfredo Gomez, Benoit Fonck, André Ayoun and Gianni Inzerillo

Abstract The SESAR 1 programme, which aims at improving Air Traffic Management in Europe, is entering into its seventh and last year of life. After two years, it reached a cruise regime, with simplified System Engineering and Management processes, to deliver each year a set of "SESAR Solutions", ready for industrialisation. Along with the management of the ongoing Program, the SESAR Joint Undertaking (SJU) prepared a following programme SESAR 2020, to continue the development and validation of next improvements organized to benefit from the lessons learnt during SESAR 1:

- Coarser granularity,
- Better and more integrated "strategic information" management,
- More systematic management of System Engineering data, including Requirements, Validation Objectives, Validation results and Architecture models,
- Strict monitoring of maturity progress of each SESAR Solution with predefined maturity criteria,
- Focusing on annual Releases of mature Solutions.

These lessons learnt can be applied to any System of Systems Research and Development programme, where coordinated System Engineering is a key issue.

For more information on SESAR, visit www.sesarju.eu or contact communications@sesarju.eu.

A. Gomez (✉) · B. Fonck
SESAR Joint Undertaking (SJU), Brussels, Belgium
e-mail: alfredo.gomez@sesarju.eu

B. Fonck
e-mail: benoit.fonck@sesarju.eu

A. Ayoun · G. Inzerillo
AIRBUS Defence and Space, SESAR Industrial Support, Brussels, Belgium
e-mail: andre.ayoun@airbus.com

G. Inzerillo
e-mail: gianni.inzerillo@airbus.com

G. Auvray et al. (eds.), *Complex Systems Design & Management*,
DOI 10.1007/978-3-319-26109-6_1

1 Introduction

1.1 SESAR Programme Objectives

The Single European Sky Air Traffic Management Research and Development (SESAR) Programme aims to modernise the Air Traffic Management (ATM) in Europe and represents the technological pillar of the Single European Sky. The programme has been launched in 2009 for a 8 years duration, and is partly funded by the European Commission. To manage the programme, the SESAR Joint Undertaking (SJU) has been created to ensure the modernisation of the European Air Traffic Management through the coordination and concentration of all relevant research and development efforts. The SJU is a public-private partnership, established by the European Union together with EUROCONTROL and involves the active participation of European and non-European ATM industry stakeholders (Fig. 1).

As a "System of Systems" (SoS), Air Traffic Management (ATM) in Europe will be improved by simultaneous and coordinated evolutions of its constituting systems. The SESAR R&D Programme aims at reaching an ambitious performance target by developing a large collection of "Operational Improvement Steps" (OIs). This development is achieved by more than 200 projects, themselves involving a number of partners working at their own site throughout Europe.

The current phase addresses the Research and Development (R&D) activities to define the Operational concept and technical solutions to meet the challenging performance targets for 2020:

- 27 % increase in Europe's airspace capacity,
- 40 % reduction in accident risk per flight hour despite air traffic increase,

Fig. 1 SESAR actors

- 2.8 % reduction per flight in environmental impact (e.g. CO_2 emission),
- 6 % reduction in cost per flight.

The SESAR program deals with a collection of pre-identified Operational Improvement steps (OIs) and corresponding Enablers (ENs) that need to be matured in two ways:

- Refinement of their definition,
- Verification and validation (V&V) aiming at increasing the confidence in their feasibility and ability to achieve the requirements, including allocated performance requirements.

The R&D activities are achieved by a high number of entities (Air National or International Service Providers and Industrials) that have their own methods, interests, and programme of work but share the common goal to integrate the validated improvements in their operational environment and products.

1.2 System Engineering Management in SESAR 1

The SESAR 1 programme addresses the R&D phases, up to the level V3 of the E-OCVM maturity scale (see maturity scale in 9 and Table 2) of the ATM System of Systems (SoS). A system of systems is "a set or arrangement of systems that results when independent and useful systems are integrated into a larger system that delivers unique capabilities […] SoS systems engineering deals with planning, analysing, organizing and integrating the capabilities of a mix of existing and new development systems into an SoS capability greater than the sum of the capabilities of the constituent parts" [5]. According to the SoS categories as proposed in 6 and 7, see Table 1, the SESAR SoS could be considered as relevant of the so-called "acknowledged" category, because the contributing systems have their own life-cycles and independent Management and Technical Authorities.

During such a Research and Development phase, the various concepts and corresponding solutions are at different stages of maturity and their development and validation need to be at minimum coordinated, and, to some extent, managed. In a previous paper (8), the authors presented the issues related to simultaneously direct the programme in a top-down performance-driven approach, and build the solutions in a bottom-up collaborative and concurrent fashion.

This situation led to adopt a customized System Engineering Management approach, with the following main features:

- Integration of the System Engineering Management Plan (SEMP) in the Programme Management Plan: the key System Engineering (SE) processes have to be simple and applied by a variety of actors having heterogeneous SE backgrounds,

Table 1 SoS categories

Type	Definition
Virtual	Virtual SoS lack a central management authority and a centrally agreed-upon-purpose for the system-of-systems. Large-scale behaviour emerges—and may be desirable—but this type of SoS must rely on relatively invisible mechanisms to maintain it
Collaborative	In collaborative SoS, the component systems interact more or less voluntarily to fulfil agreed upon central purposes. The internet is a collaborative system. The internet engineering task force works out standards but has no power to enforce them. The central players collectively decide how to provide or deny service, thereby providing some means of enforcing and maintaining standards
Acknowledged	Acknowledged SoS have recognized objectives, a designated manager, and resources. However, the constituent systems retain their independent ownership, objectives, funding, development, and sustainment approaches. Changes in the systems are based on collaboration between the SoS and the system
Directed	Directed SoS are those in which the integrated system-of-systems is built and managed to fulfil specific purposes. It is centrally managed during long-term operation to continue to fulfil those purposes as well as any new ones the system owners might wish to address. The component systems maintain an ability to operate independently, but their normal operational mode is subordinated to the central managed purpose

- Avoiding micro-management and focussing on main outcomes. Focus management effort on high level milestones and give particular visibility to the most mature concepts and solutions, having reached the V3-maturity (see maturity scale in 9 and Table 2), during the last year: an annual "Release" approach has been adopted.
- Synchronizing the various cycles to ensure technical coherence of the overall reference information. During the annual cycle, the Release Strategy, Integrated Roadmap and Verification and Validation Roadmap are aligned at predefined times during the year, and related information is configuration-managed.

2 Lessons Learnt from the SESAR 1 Programme

As, from 2015 onwards, the SESAR Programme is extended in the framework of Horizon2020, to address further improvements and prepare to the deployment of SESAR solutions, the full programme structure as well as the management and system engineering principles have been reconsidered to deal more efficiently with the difficulties met in SESAR 1. This Programme extension is called SESAR2020 (S2020).

Table 2 E-OCVM Maturity Scale

Phase	Objective	Activities
V0	ATM Needs	Establish and quantify the need and drivers for change. The current and potential future situation should be analysed and the improvement areas and objectives identified and performance targets established
V1	Scope	Identify the operational/technical solutions for meeting the target performance identified in phase V0. The proposed operational concept(s) and associated technical solution(s) should be defined in sufficient level of detail to enable the establishment of an appropriate performance/assessment framework, the identification of potential benefit mechanisms, scope of potential applicability and initial cost estimates (order of magnitude) to justify R&D. …
V2	Feasibility	Develop and explore the individual concept elements and supporting enablers until the retained concept(s) can be considered operationally feasible or it can be established that further development is no longer justified. … This stage will mainly establish the feasibility from the operational and transitional view point and provide initial elements for technical feasibility
V3	Pre-industrial development and integration	– Further develop and refine operational concepts and supporting enablers to prepare their transition from research to an operational environment; – Validate that all concurrently developed concepts and supporting enablers (procedures, technology and human performance aspects) can work coherently together and are capable of delivering the required benefits; – Establish that the concurrent packages can be integrated into the target ATM system
V4	Industrialisation	Applicable specifications are submitted, approved and published as Standards by the ATM
V5	Deployment	The supply industry builds, installs, integrates and validates on-site systems/facilities/infrastructure
V6	Operations	Users and service providers operate in accordance with the deployed concepts and supporting enablers
V7	Decommissioning	Plan and execute the termination of the use of concepts and enablers by users and service providers. …

Based on the lessons learnt in SESAR 1, the main changes in the management of S2020 are:

- Granularity,
- Strategic information management,
- System Engineering data management,
- Maturity monitoring,
- Release approach for delivering SESAR Solutions.

2.1 Granularity

Whilst the SJU had to manage more than 200 domain-oriented Projects in SESAR1, resulting in a heavy coherence coordination workload, the new SESAR2020 programme is composed of 19 (and bigger) Solution-oriented projects and 4 main Transverse projects. This approach relies on the subsidiarity principle and avoids "micro-management" overload.

The new Projects in S2020 address both Operational and System aspects: this should alleviate the workload of coordinating Operational and Technical projects to ensure that the Technical Requirements are derived (trace to) Operational Requirements, to ensure that the prototypes and Validation Platforms are built to implement the functionalities and features necessary to validate the corresponding operational concepts. Now, these coordination tasks will stay internal to the project, with less cumbersome coherence maintenance activity.

Each solution-oriented project is in charge of developing and validating a set of SESAR Solutions, each one being built on one (or several) Operational Improvement step(s) and on supporting Technical enablers. In addition to solution projects, transverse projects are implemented to ensure overall coherence and to ensure the alignment of the validation methods with the directions and guidance in transverse areas: Master Planning, Content Integration, Support to Verification and Validation.

2.2 Strategic Information Management

The SESAR 1 programme shew the need for managing "strategic information", composed of:

- Scope and Initial Operational Capability (IOC) target date of Operational and Technical Improvements: the corresponding data populate the ATM Master Plan. These data are regularly updated to take into account Change Request, mostly to refine the definition and the scope of future Operational Improvements,
- Target dates for each Operational Improvement step (OIs) to reach the successive maturity levels (V1, V2, V3): these data compose the Release Strategy. The Release Strategy provides a planning view indicating the target Release for a SESAR Solution's V3 maturity. The Release Strategy also describes the maturity currently achieved and the maturity expected to be reached at specific points in time. The Release strategy is periodically updated to take into account the progress of the development and validation.
- Organizational and Programmatic data: the OIs are grouped into Operational Focus Areas (OFA). Projects are also associated to OFAs. The Projects also prepare and run Validation Exercises to validate SESAR solutions. Correspondence between projects, OIs and SESAR Solutions and Exercises need to be maintained and configuration managed.

During SESAR 1, the three above mentioned threads were managed separately in configuration, with separate control boards. Change impact assessment was mostly controlled at project level, rather than at programme level. So, there is a need to better manage and structure strategic information, in a comprehensive and shared data base strictly managed in configuration, to support alignment of the various elements of each solution within projects and across interdependent projects.

2.3 System Engineering Data Management

System Engineering data are interrelated data consisting in:

- High level performance targets,
- Operational and System Requirements,
- Validation Objectives,
- Validation results,
- Architecture models.

In the SESAR 1 programme, Operational and Technical Requirements have been captured from a significant amount of project deliverables to create a consolidated and coherent Requirements data base, and to ensure the traceability from high level requirements down to technical requirements for supporting systems. To automatically capture these requirements, it has been necessary to define strict document templates with appropriate fields. The "import"(in the DOORS™ data base) procedure required that deliverables were compliant with a required format. In the beginning of the SESAR programme, many documents did not comply with the format, resulting in a lack of completeness of the requirements data base. The situation has progressively improved but there are still documents delivered by projects, which do not comply with the required format.

In the same way, Validation objectives, defined planning a validation roadmap and then allocated to validation exercises, are captured in the Requirements data base. Indeed, Validation objectives trace to Requirements. After Validation exercises execution, validation results support marking validation objectives as fulfilled or not met. Overall traceability analysis then permits to compute validation coverage.

In practice, the traceability chain was often broken, due to the insufficient "importability" of SE data from some deliverables (and on the resulting overload associated to quality check). The coverage analysis has been therefore restricted to the coverage of the Validation Objectives by Validation Results for the SESAR Solutions validated within the current release.

The situation for Architecture models was comparable: at the beginning, the target was to ensure consistency between project deliverables content and architecture models in a central repository. Coherence between architecture models and Requirements was also targeted, but quickly given up, due to difficulty to both obtain the data, and ensuring that they refer to the same configuration.

This experience reinforces the belief that there is a need to better collate and manage all SE data in a controlled fashion, with centralised configuration and change management. Centralised repositories will ensure the overall coherence of the SE data elaborated by various projects. SE data should be formally considered only when they achieve a sufficient quality.

To ensure the consistency, a driver is the minimization of translation operations: one option is to allow the Project contributors to directly enter data in a centralized data base. Such a data centric approach presents the advantage of ensuring coherence between Operational concept documents with architecture models and requirements within the scope of the Project. The reference is the central repository and not the document. The document is generated from the repository. However, it necessitates a coherence control during the import: the data need to be consistent with the strategic reference information (overall configuration management): several feedback loops may be necessary to align the models and data at high level with all supporting contributions. Another difficulty is the need to have common tools, or common access to a central tool, at all project sites and to ensure that writing rules are known and applied.

For these reasons, variants for data centric approaches are still under study (Fig. 2).

2.4 Maturity Monitoring

Maturity monitoring is one of the main positive lessons from SESAR 1: during the programme performance, the Maturity criteria (for each maturity step V1, V2, V3) have been refined and objectives metrics to assess them have been developed. This resulted in a well-accepted tool to support both self-assessment (by project managers) and assessment by SE reviewers during the 3 SE reviews performed in the Release process. Maturity reports are produced twice a year, which give an overall view of the programme solutions (more precisely OIs) maturity. This monitoring is fundamental to prepare transition to Industrialisation and Deployment, but is also of upmost utility to re-plan development and validation activities, when necessary.

So, the SE reviews will be replaced by Maturity gates, to monitor the maturity progress of each SESAR Solution (operational or technological solution), for the following transitions:

- Exploratory Research to Industrial Research
- V1 to V2, to assess if the solution is sufficiently clear to be further developed,
- V2 to V3, to assess its feasibility and expected benefits, based on performance first results,
- V3 to V4, to assess its readiness for industrialisation and further deployment,
- V3+ to V4: Very Large Scale Demonstrations to de-risk the deployment.

The maturity of the SESAR Solutions is assessed along seven maturity threads:

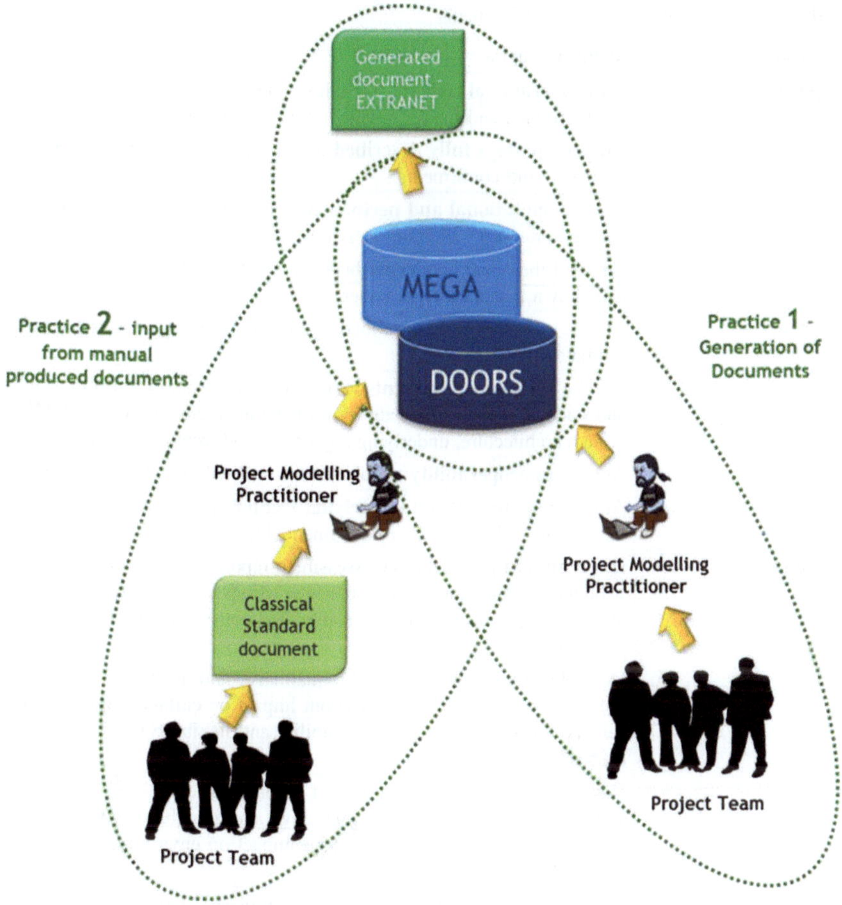

Fig. 2 Two practices for SE data capture: data-centric (*1*) and document-centric (*2*)

- Operations: accuracy and relevance of the Operational concept description,
- System: technical solution is suitable to support the operational concept,
- Performance: in all Key Performance Areas, the performance is assessed and meets the target,
- Standards and Regulations: the standardisation and regulation processes are addressed,
- Transition: transition between current and future methods is addressed and satisfactory,
- Programme: programme aspects, in particular coherence and synchronization of related developments are addressed,
- Validation: validation results demonstrate the maturity, with appropriate validation methods.

The Table 3 illustrates the criteria associated to the V3 maturity level:

Table 3 V3 maturity criteria in the 7 threads

Thread	Maturity criteria
Operations	Is the **operational concept** (including roles, working methods, training needs) refined and further detailed after V3 activities and documented?
	Are the **OI steps fully described** and documented e.g. IOC-dates estimated and confirmed, …
	Are the **operational and performance requirements** stable and updated after V3 validation activities?
	Have all the related **concepts been integrated** and validated together, and shown that they work coherently?
System	Are the **enablers** fully described and documented e.g. IOC-dates estimated and confirmed?
	Are the **system requirements** verified on a verification platform, stable and updated after V3? Verification of the integrated prototype, HMI, system architecture, underlying algorithms and technology is successful
	Are the **interoperability requirements** updated after V3 activities?
	Are the requirements on **underlying technology** (e.g. communication, navigation and surveillance) documented?
Performance	Has a **human performance** assessment report been completed? Do validation results confirm that the interactions between human and technology are operationally feasible, and consistent with human performance requirements?
	Do validation results confirm the qualitative and quantitative evidence obtained in previous V phases about impact on **capacity, quality of service KPAs** (efficiency, predictability and flexibility) and **cost-effectiveness**?
	Has an **environmental Assessment** report been completed? Do validation results confirm the qualitative and quantitative evidence obtained in previous V phases about impact on environmental sustainability?
	Has a **safety assessment** report been completed? Do validation results confirm the qualitative and quantitative evidence obtained in previous V phases about impact on safety?
	Has a **security assessment** report been completed? Do validation results confirm the qualitative and quantitative evidence obtained in previous V phases about impact on security?
	Are the assessments results **in line with what is targeted** for that concept? In case of deviation, has been the impact on the overall strategic performance objectives/targets analysed?
	Has the V2 **cost estimation** associated to the OI steps and associated enablers been updated and refined per deployment scenario and stakeholder after Validation activities in V3?
Standards and Regulations	Is the material produced sufficiently developed and mature to support the development or update of **operational and technical standards** in V4?
	Is the material produced sufficiently developed and mature to support the **regulation** process in V4?

(continued)

Table 3 (continued)

Thread	Maturity criteria
Transition	Has the **transition** analysis been refined by taking into account evolution of the operational concept and supporting enablers during V3 phase?
	Are there **recommendations** proposed for V4?
Programme	Is there evidence that other **related OI steps and enablers** are at the expected level of maturity?
Validation	Do **validation** results confirm the **quantitative and qualitative evidence** on the operability and technical feasibility obtained in previous V phases?
	Were the V3 validation exercises executed in an operational **environment representative of the target deployment scenario**?
	Were the V3 validation activities executed using a **validation technique suitable** for that maturity level
	Are reference, solution scenarios and most relevant **non-nominal situations** considered in the validation?
	Are the addressed **validation objectives coherent** with the validation strategy and with the expectations in V3?
	Has the industry based **platform** been successfully verified and accepted prior to the validation activity?

This assessment is based on the results of validation activities, the status of standardisation and certification processes and an assessment of interoperability. While E-OCVM will continue to be applied, the link with Technology Readiness Levels (TRLs) will also be ensured. TRL is a well-known maturity scale (initially proposed by the US Department of Defence and by NASA, now normalised by ISO ref. 16290:2013), suited to communicate achievements. The Fig. 3 provides a mapping between Technology Readiness Levels (TRLs) and E-OCVM maturity levels and shows at when the maturity Gates take place.

2.5 *Release Approach for Delivering SESAR Solutions*

The Programme and Release Delivery lifecycle is an annual sequence of Programme Milestones, associated to key outputs of Transversal activities which have an impact on solution-oriented projects. In the SESAR 1 programme, particular management focus was given of solutions targeting their V3 maturity level in the coming year "n". So, the corresponding set of V3-validated solutions constitutes the "Release n". The release process has demonstrated to be a very important and positive feature of the SESAR 1 Programme, to federate energies around visible objectives (Fig. 4).

The annual Release Approach will be kept in SESAR 2020, to ensure that the ATM Master Plan is followed and to coordinate efficiently with the authorities in

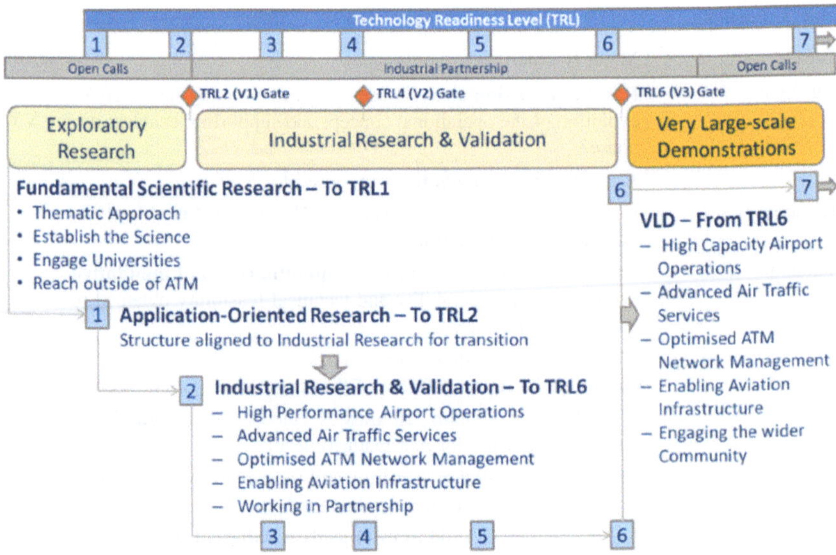

Fig. 3 SESAR maturity phases (TRL vs. E-OCVM)

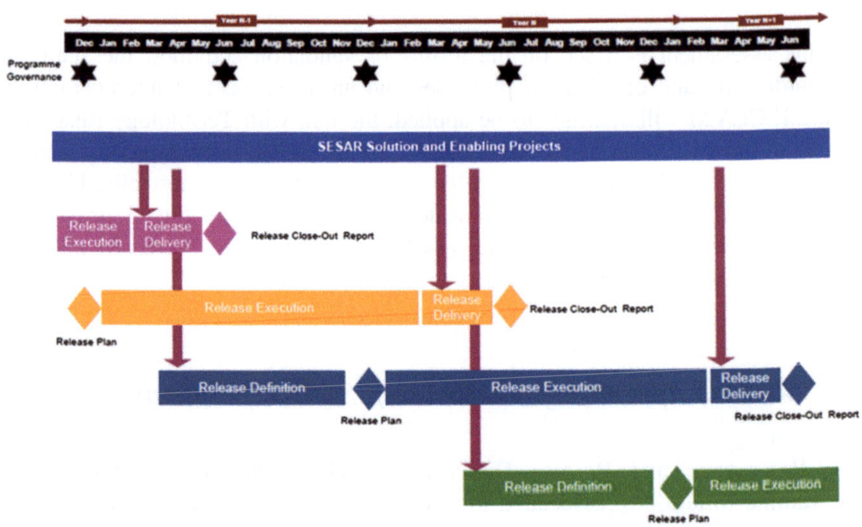

Fig. 4 Overlapping release life cycles and relationship with solution projects

charge of deployment of the solutions, namely the "Deployment Manager" entity. In essence, the selection criteria to enter a Release are:

– The solution has demonstrated to have already reached the V2 maturity level at a preceding V2 maturity Gate,

- The solution project has provided a validation plan giving confidence in its ability to complete the V3 validation activities within the coming year n; the validation plan should include the next V3 maturity gate and should show that dependencies with all contributing projects have been agreed.

3 Applicability to Other Large Scale Systems

The lessons learnt, in both fields of Programme Management and System Engineering methodologies, can be generalized to many categories of large scale systems and in almost all systems of systems, but with variable importance, depending on their specific features:

- Granularity: selecting the proper granularity for defining projects is, at first glance, a management issue but is closely related to the system engineering approach (lifecycle, interface management, V&V approach in relation with the architecture, requirements management, etc.). It is related to the Work Breakdown Structure (WBS) and to the Organizational Breakdown Structure (OBS). On one hand the project size has to be small enough to ensure that management and governance instances have visibility on tasks progress to favour earned value management (EVM) and fine monitoring. On the other hand the management overhead should be limited with respect to the need for technical competences to control the overall consistency. The lesson learnt from SESAR is that an early-defined small granularity results not only in a high management and monitoring workload, but also in a risk of missing the right organisation. The best project definition and organisation is the result of a trade-off between the need for self-consistent projects, with reduced inter-programme interfaces, and the need for transparency on technical and performance progress.
- Strategic information management focuses on essential elements that support the programme baseline. This process is also relevant of both Programme Management and SE management, for it addresses programmatic information (IOC dates, even major maturity milestones/target dates) as well as the performance targets and the high level content of the solutions that contribute to the expected capabilities. In usual System Engineering management, this is addressed within the high level programme management (key milestones) and within SE management as high level requirements that can be derived and allocated to subsystems. In the case of the R&D of a SoS, the need is more to ensure that the independently conceived solutions contribute to the overall targeted capability and performance (within Key Performance Areas) rather than a top-down derivation. High visibility should be given to this strategic information, which has to be shared by the partners. This allows better adherence to the common goal, and better understanding of the priorities.

- System Engineering data management is a standard (ISO IEC/IEEE 15288 norm) recommended for any system design. The concern in the design, integration and validation of relatively independent systems of systems is that each authority has its own SE management approach and tooling, and moreover, has often serious reasons not to disclose details that could hinder their position (intellectual property, internal difficulties …). So, obtaining from the contributing projects the detailed SE data, with sufficient content and quality is a challenge. However, to justify the design and to monitor the validation coverage, it is highly important to capture the links from technical requirements to operational ones and to assign validation objectives to them. The management to a proper selection of SE data is key to monitor SoS developments: indeed, sharing the proper level of requirements is an efficient way to share the minimum needed SE information, without disclosing the technical details that need to stay proprietary and undisclosed. So, the best trade-off between detailed SE data management and only high level information data sharing depend on the phase of the project and on the competitive situation.
- Maturity monitoring is usually performed at the main Design Reviews: in the case of SoS, with many elementary solutions developments, the maturity monitoring should be performed in a more continuous way and not wait for overall design reviews. Maturity is generally well understood for technological solutions (using the TRL scale), but other criteria are not always well formalized (regulatory aspects, performance confidence, validation coverage, etc.). Based on SESAR experience, generic maturity criteria could be applied to a large variety of SoS. These criteria permit to quantitatively assess the maturity, based on validation results (SE data) and other provided evidence.
- Release approach for delivering solutions: in a large SoS, where high level capabilities are built by parallel and asynchronous development of many elementary improvements, the risk is to let the research drift. A good way to motivate all actors to delivering visible validated solutions is to define every year the content of the next year release and to obtain commitment of all implied actors in the validation of the their respective solutions. Such a Release approach helps to focus resources on identifying and monitoring the most mature concepts, candidate to delivery. This good practice is not much addressed in SE reference documentation.

4 Conclusion

The SESAR2020 organisation and processes will benefit from lessons learnt in SESAR 1. Such lessons are more generally applicable to any large SoS development, especially in the R&D phases. To summarize, the main recommendations are:

- Organize the programme in large size projects dealing with relatively independent solutions. Coarse granularity and delegation of management are necessary to ensure proper steering and control of the overall programme,
- Structure and manage integrated strategic information, which has to be shared by all actors to build coherent and synchronized solutions,
- Centralize System Engineering data, including Requirements, Validation Objectives, Validation results and Architecture models, to ensure overall consistency,
- Monitor strictly Maturity progression at maturity gates, with predefined maturity criteria,
- Give attention and visibility to solutions that are to be validated in the coming year, by using an annual Release Approach.

References

1. Systems Engineering Handbook, a guide for system lifecycle processes and activities, International Council on Systems Engineering (INCOSE), version 3.1 Aug 2007
2. Baldwin, K.: Systems of systems: challenges for systems engineering. INCOSE SoS SE Panel, 28 June 2007
3. ISO/IEC 15288: Systems engineering—system life cycle processes (2002)
4. Dahmann, J., Baldwin, K.: Understanding the current state of US defense systems of systems and the implications for systems engineering. IEEE Systems Conference, Montreal, Canada, 7–10 April 2008
5. Department of Defense: System of systems engineering. Defense Acquisition Guidebook, Washington, DC, 14 Oct 2004
6. Maier, M.: Architecting principles for systems-of-systems. Sys. Eng. **1**(4), 267–284 (1998)
7. Gomez, A., Fonck, B., Ayoun, A., Inzerillo, G: Concurrent system engineering in air traffic management: steering the SESAR programme. CSD&M 2013, Paris
8. European Operational Concept Validation Methodology (E-OCVM), version 3, vol. I, Feb. 2010, Eurocontrol

Co-Engineering: A Key-Lever of Efficiency for Complex and Adaptive Systems, Throughout Their Life Cycle

Anne Sigogne, Odile Mornas, Edmond Tonnellier and Jean-Luc Garnier

Abstract Thales Group designs, develops, produces, supports, operates innovative solutions in large and various domains (Aerospace, Space, Defence, Aerospace, Ground Transportation, Security, etc.) where the operational performances are more and more critical. In this context, to ensure competitiveness and remain leader on the market, Thales has investigated in an extension of the recommended Integrated Product and Process Development approach (see [DoD IPPD], [INCOSE SE HB], [CMMI]), applied for Co-Development towards a "Co-Engineering approach" addressing all stages and concerns of the operational system as a key lever of efficiency and SE benefits achievement. This paper presents the implementation in Thales of this Co-Engineering approach identifying major principles to be mutually agreed and applied (on Technical and Organisational aspects) per System Life Cycle stage, necessary changes to be led, and finally, an illustration by typical scenarios as Returns of Experience.

A. Sigogne (✉)
Thales Global Services, 19/21 avenue Morane Saulnier, 78140 Vélizy-Villacoublay, France
e-mail: anne.sigogne@thalesgroup.com

O. Mornas
Thales Université, 67 rue Charles de Gaulle, 78350 Jouy-En-Josas, France
e-mail: odile.mornas@thalesgroup.com

E. Tonnellier
Thales Systèmes Aéroportés, 2 Avenue Gay Lussac, 78851 Elancourt Cedex, France
e-mail: edmond.tonnellier@fr.thalesgroup.com

J.-L. Garnier
Thales Technical Directorate, 1 avenue Augustin Fresnel, 91767 Palaiseau cedex, France
e-mail: jean-luc.garnier@thalesgroup.com

G. Auvray et al. (eds.), *Complex Systems Design & Management*,
DOI 10.1007/978-3-319-26109-6_2

1 Introduction

"Why Co-Engineering"? The genesis of this approach is linked to the main features of Thales Group and to the new challenges that the group has to face to.

Co-Engineering addresses both Collaborative and Concurrent Engineering. It impacts both organisation but also team mind-set sharing objective and system vision, and so achieving Systems Engineering benefits.

Co-Engineering approach contributes to satisfy one main Customer expectation: optimize the system Life Cycle Cost, a main concern for complex and adaptive systems notably when life time extends over several decades.

Note: Any engineering artefact (product, service, piece of software or hardware) is discussed in this document with a systemic approach and is called a "system".

1.1 The Main Features of Thales Group

Thales Group is a "key player" in various domains (Aerospace, Space, Ground transportation, Defence and Security) aiming to promulgate "**A collective intelligence for a safer word**" (Fig. 1).

This collective intelligence is based on a word wide implementation of employees, supported by recurrent investments.

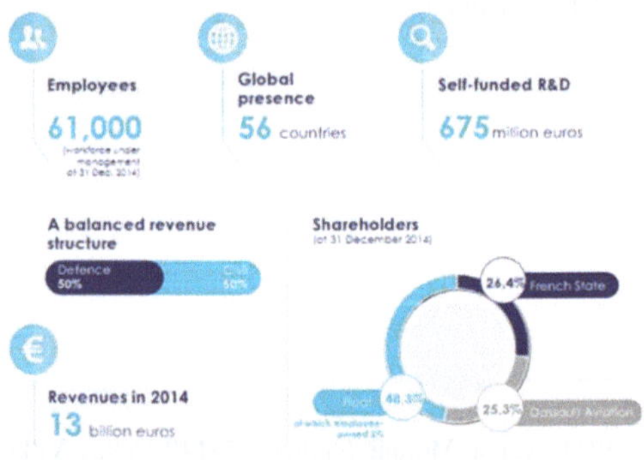

Thales is precisely involved in following domains (Fig. 2).

Thales commits itself in various engineering activities all along the **System Life cycle stages** with an increasing effort in after-development stages. The scope must address all stakeholders so encompass both the **System of Interest** and the necessary **Enabling Systems** that interoperate (Figs. 3 and 4).

Fig. 1 The Thales Group positioning on markets

Fig. 2 The diversity of domains addressed by Thales Group

1.2 The Genesis of Co-Engineering Within Thales

To remain competitive whilst maintaining Customer satisfaction, focus has been made on a necessary Engineering "Collective value", aiming to carry an "efficiency" mind-set within stakeholders beyond the Integrated Product and Process Development initiative (see [DoD IPPD], [INCOSE SE HB], [CMMI]) which is efficient for small teams with one scope of responsibility.

The objective is to reach the real challenge for both Systems providers, Acquirers and Users: **"Optimise the Life Cycle Cost"** (as illustrated in Fig. 5, from [INCOSE SE HB]) for an observable and compliant "**operational performance**"

The "**Co-Engineering**" approach is an answer to this challenge.

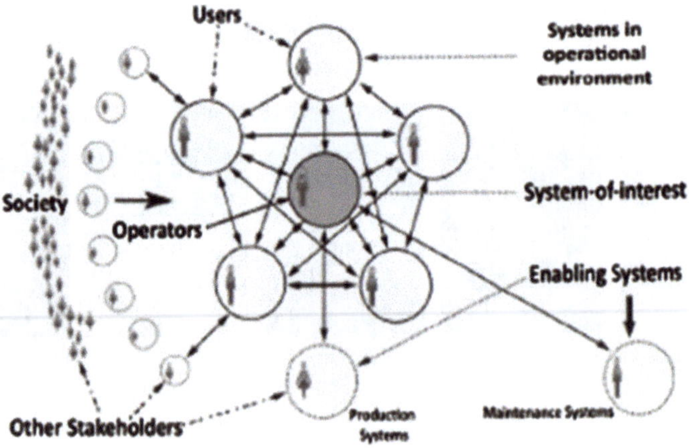

Fig. 3 The system of Interest and its typical stakeholders

Life cycle stages	Purpose
Exploratory Research	Identify stakeholder's needs Explore ideas and Technologies
Concept	Refine stakeholder's needs Explore feasible concepts Propose viable solutions
Development	Refine system requirements Create solution description Build system Verify and validate system
Production	Product system Inspect and verify
Utilization	Operate system to satisfy user's needs
Support	Provide sustained system capability
Retirement	Store, archive or dispose of the system

Fig. 4 Example of system life cycle stages ([INCOSE SE HB])

As foreseeable, "Systems engineering" being basically an **inter-disciplinary** approach that has to consider the **complete problem** (i.e. operations, cost and schedule, performance, training and support, test, manufacturing and disposal) has been delegated to define, implement and experiment this approach then secure its large deployment, especially for "Concept" to "Support (including Services)" stages.

But the drawback of this inter-disciplinary approach is that Systems Engineering could be seen as an upper layer over disciplines and specialties working in silos to deliver systems parts. In that case, each team concentrates his effort on his objectives and may miss the global ones.

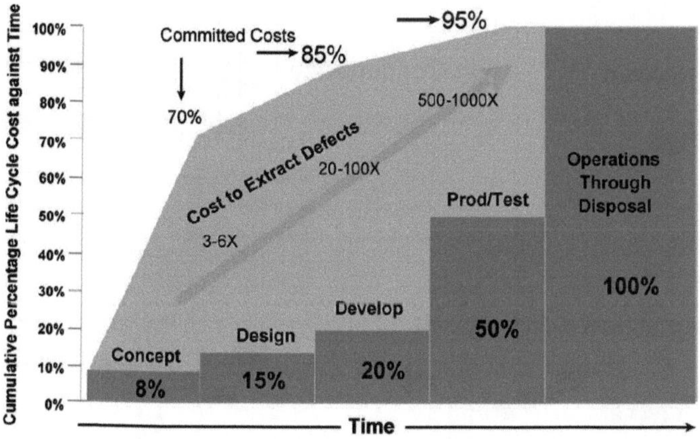

Fig. 5 Early commitments against through life cycle stages costs

2 Co-Engineering Definition and Principles

2.1 The Stakes

To face more and more tightened time constraints, engineering teams involved in a project **work in parallel, rather than in a sequential way** (concurrent Engineering). This is not sufficient to satisfy the requirements in terms of efficiency and competitiveness: the Co-Engineering intends to **improve and increase the collaboration** between these teams.

Today, Systems Engineering of complex systems follows various Engineering development cycles: vee model, spiral model, incremental and iterative model, etc. [see Das V-Modell, August 1992, Spiral model (Boehm 2000)]. As illustrated by following Fig. 6, these models could generate a lack of "global vision" due to "too focused" concerns (deliverables, cost, etc.) and so a lack of coherency between disciplines.

Co-Engineering is based on a **shared vision** of the problem and project outcomes, with common objectives clearly defined in order to get this coherency.

This shared vision should cover **all the stages of the system life cycle**, from Concept up to the Retirement of the system, and involves **all the stakeholders of the system** (e.g. customer, partners and suppliers, Bid and Project management, purchases, product policy, engineering including production and service).

The global objective is to optimize the **way of working** between engineering disciplines, specialties, manufacturing and services and avoid cascading of analysis and loops with rework too often induced by "**silos**" effect.

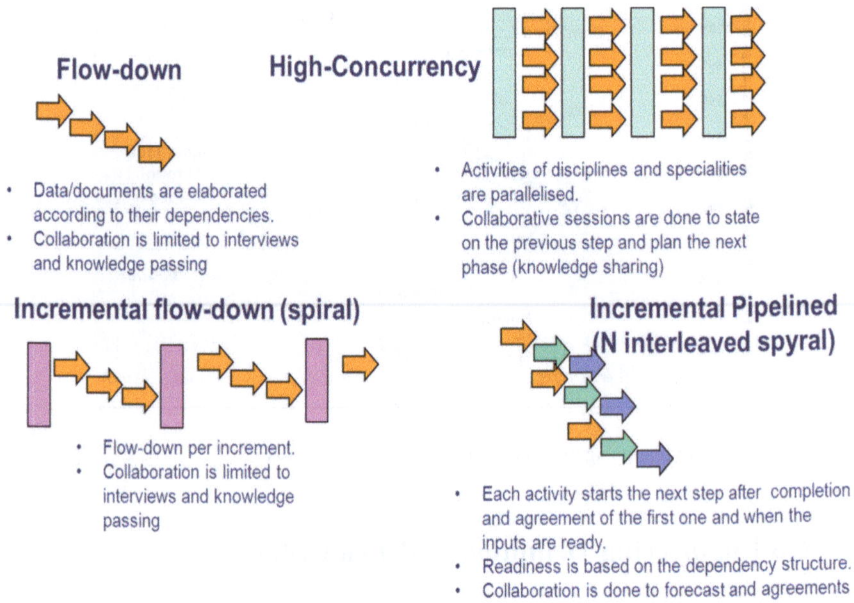

Fig. 6 Typical engineering products development life cycles

2.2 *Basic Definitions*

Co-Engineering corresponds to "***Concurrent*** *engineering activities performed in a* ***collaborative*** *and* ***cooperative*** *way. Based on a* ***shared vision*** *of the scope of the solution, the actors jointly make analyses, decide and master risks for* ***collective value*** *enhancement*".

Where followings terms must be understood as:

- **Concurrent engineering**:
  The application of multiple engineering disciplines to perform allocated activities in several different but related areas at the same time so the activities are coordinated and mutually supportive.
- **Collaboration**:
  Collaboration means working together with shared dynamic goals to achieve collective results that benefit to all parties involved. It implies a higher degree of commitment, mutual trust, and sense of belonging and common interest than cooperation.

A critical feature of collaboration is vision sharing by all the stakeholders involved in producing targeted results in complex contexts.

- **Cooperation**:
  Mutual agreement to work on consolidated artefacts conformed to a set of individual objectives

2.3 Main Principles

Multi-points of view approach
Following Fig. 7 aims at representing the typical System life cycle activities so as to facilitate the identification of the essential points of view which must be captured, understood, and confronted to ensure the efficiency of the Co-Engineering approach.

Shared vision concept
The purpose of creating a shared vision is to achieve a unity of purpose regarding the Systems Life activities with respect to the contract limitations and the scope of responsibility.

The value of a shared vision is that people understand and can adopt its principles to guide their actions and decisions. Shared visions tend to focus on an end state while leaving room for personal and team innovation, creativity, and enthusiasm.

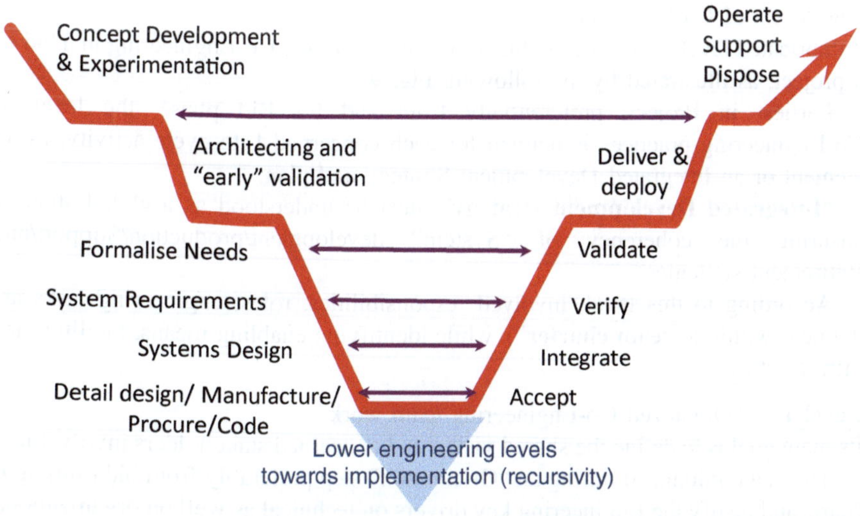

Fig. 7 System life cycle activities concerned by Co-Engineering

The activities of the individuals, teams, and project are aligned with the shared vision (i.e. the activities contribute to the achievement of the objectives expressed in the shared vision).

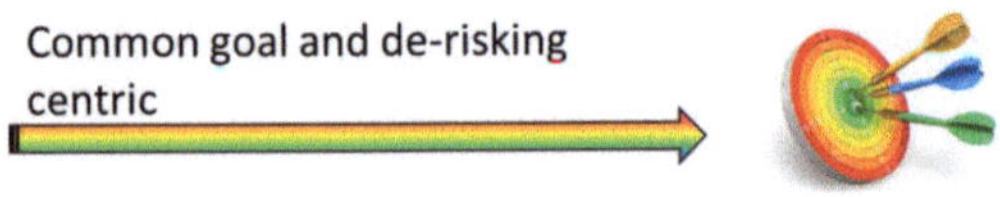

Initiated by the **Bid team** (or equivalent to launch a "system" development) and consolidated all along project activities as relevant, the shared vision shall propose a global answer to the following questions:

- What is the target in terms of context, contents, and constraints?
- Who is involved for which activity in terms of stakeholder representatives?
- When do activities require Co-Engineering approach to be optimally performed?
- Where does Co-Engineering take place?
- What is the level of collaboration necessary for what activity? What are the adequate means required for the activity?

It is important to ensure that the shared vision is **understood by all stakeholders** involved in the "System" life cycle in a comprehensive and homogeneous way (typically, avoid the well-known divergence of points of view on "swing" concept, illustrated humorously by the following Fig. 8).

Creating a shared vision requires that all involved people in the project have an opportunity to speak and be heard about what really matters to them. The project's shared vision captures the project's guiding principles, including mission, objectives, expected behaviour, and values. Technics of **team building** may be used to raise confidence.

The levels of Co-Engineering

4 imbricated levels have been defined to characterize the Co-Engineering in a bid or a project, as illustrated by the following Fig. 9.

Earliest in Project, preferentially from and for Bid phase, the level of Co-Engineering practices is defined for each concerned Life cycle activity, as an element of an Integrated Development Strategy and plan.

"**Integrated Development strategy**" must be understood as a global strategy ensuring the coherency of "System" development/production/Support/and Retirement strategies.

According to this level, involved responsibilities, roles, engineering tasks are defined (within a "**team charter**"), while identifying enabling means, facilities and infrastructure.

<u>**Level 1**</u> as "Organized Co-Engineering team work"

Its main goal is to define the shared vision and concerned stakeholders involvement.

This data, initialized during an "Orientation" step preferably from bid (aiming to frame and justify the Engineering key drivers on technical as well on organizational

Fig. 8 Typical misunderstanding of stakeholders on "swing" vision

aspects), will be updated only if changes occur in the context of the project. It should be a basis for the elaboration of the "Integrated Development Strategy".

The stakeholder involvement is detailed in the SEMP (System Engineering Management Plan) and downward in other concerned engineering subordinate plans then updated as necessary during the phases of the development cycle. The period of Co-Engineering meetings is defined depending on the main activities to perform. Extra meetings may be planned on demand.

It is important to introduce the workload due to these meetings in estimates for each stakeholder and take care to avoid overloading due to meetings.

Level 2 as "Application of the Co-Engineering method" (including level 1)
The Co-Engineering method is based on **field proven** facilities to support the expected efficiency of team work collaboration and cooperation.

During the Co-Engineering adoption phase, it is strongly recommended to take advantage of using standard collaborative facilities (Visio conference, Live meeting, etc.)

Then, all along effective Co-Engineering practice, this method shall be progressively based on identified best practices of entities (from **capitalization** process).

A particular section of the System Engineering plan may describe the planed usage of these facilities. It is fundamental to communicate as appropriate to ensure their availability at the right time.

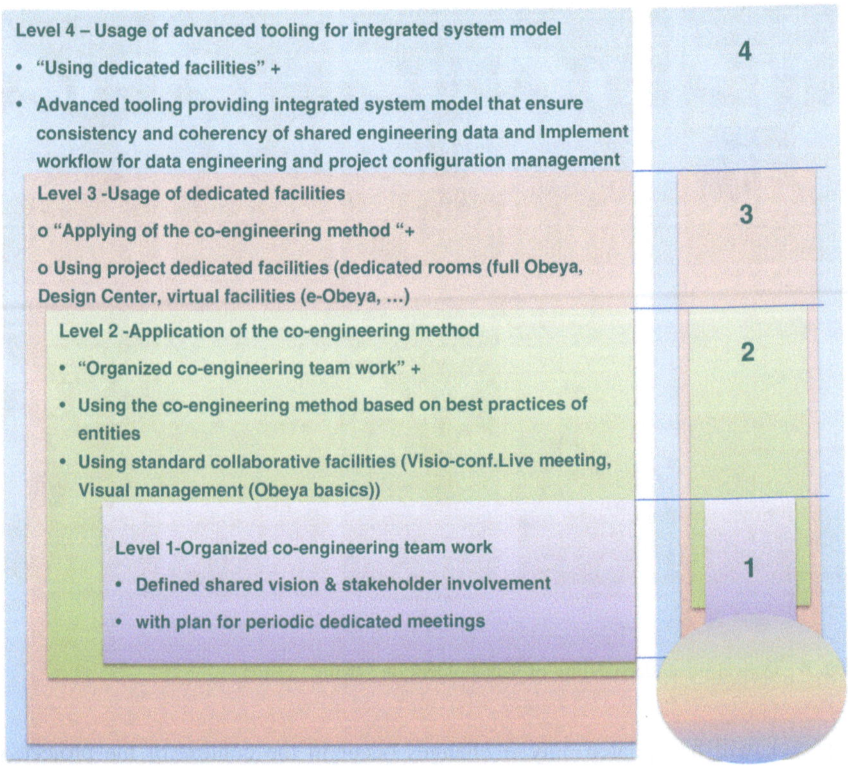

Fig. 9 The four levels of Co-Engineering defined in Thales

Practice a Co-Engineering method implies, before application in Bid and Project:

- A **Preparation** phase, before starting of concerned engineering activities, ideally earliest in bid:
  - To scope Co-Engineering practices (phase, step, engineering level(s)…)
  - To identify the stakeholders
  - To define, according the level of shared vision, information to be shared between the stake-holders
  - To define the Co-Engineering meeting types and techniques, related infrastructure to be used along the bid and project phase
  - To define the team organisation for the different working meetings
  - If facilitation is required, to identify a facilitator among the organisation
  - To adapt the organisational process as appropriate to consider the defined Co-Engineering practices
  - To put in place the management of shared Engineering data: a Model Based System Engineering (MBSE) approach being recommended.

- Then, a **Kick-Off** phase

Level 3 as "Usage of dedicated facilities" (including levels 1 and 2)
Some facilities can support and enhance the meeting efficiency. Meeting room configuration, display, space can have a significant impact on the meeting dynamics and in the participant involvement. Different types of facilities are recommended depending of the project phase and meeting objectives.

- A **creativity centre** ("creativity corner"), room or space: will help early phases, concepts and trade identification, but also problem solving. Creativity centre has to provide:
  - Room for a stand-up meeting
  - White board wall, to provide a large surface to exchange information, draft concept.
  - Large free space for people to exchange, walk around the elements.
  - Material to draft concepts (blocks, hard paper …)
- A **concurrent design centre** will provide extended engineering capabilities enabling to perform work during the session. The meeting will be around a table, the participant having to perform work on computers, to exchange data and update models. The meeting room shall facilitate the sharing of information between participants in a responsive way by ensuring to display different PCs displays on different screens. All work stations or participant lap top should be connected (multiplexer) to the different screens. This design enabling should also provide white boards to facilitate rapid graphic communication among participants or to display information, guidelines.
- An **OBEYA room** that is effective, to perform project coordination and manage actions status, solves in a short term problems. The principle is to display the key project information on the wall, to provide a global system vision, such as: baseline, risks, planning, but in particular to display the short and mid-term actions, check their progress and update or take new actions. OBEYA is based on a stand-up meeting to keep people focus on the project and to avoid distractions. OBEYA can be also implemented in a virtual way e-OBEYA using electronic boards, when the team is spread other different premises (Fig. 10).

Fig. 10 Co-Engineering dedicated facilities (creative/design center, Obeya room)

Level 4 as "Usage of advanced tooling for integrated system model" (including levels 1, 2 and 3)

As example, **Capella** provides means to ensure an engineering-wide collaboration with all stakeholders sharing the same **reference architecture**, including architects and engineers for system and subsystems, development teams, specialty engineers (e.g. interfaces design, performance, security, RAMS—Reliability Availability Maintainability and Safety—costs, mass, product line, etc.), integration and validation, customer, etc.

Capella is the provided tool that implements Thales ARCADIA (MBSE) method (related to CLARITY project): one of the main noticeable features of Arcadia is to support enterprise-wide collaboration and Co-Engineering.

Collaboration with engineering specialities is supported by **modelled engineering viewpoints** to formalise constraints and to evaluate architecture adequacy with each of them.

Collaboration with customer and subsystems engineering relies on **co-engineered models** (e.g. physical architecture), automatic initialisation of need model for sub-systems, and impact analysis means between requirements and models of different engineering levels (as illustrated by Fig. 11).

A dedicated Engineering environment, integrating as wider as possible a set of tooled-up services addressing all concerned stakeholders, is recommended for optimally enables the Co-Engineering practice at level 4, as illustrated (partially) by the following Fig. 12.

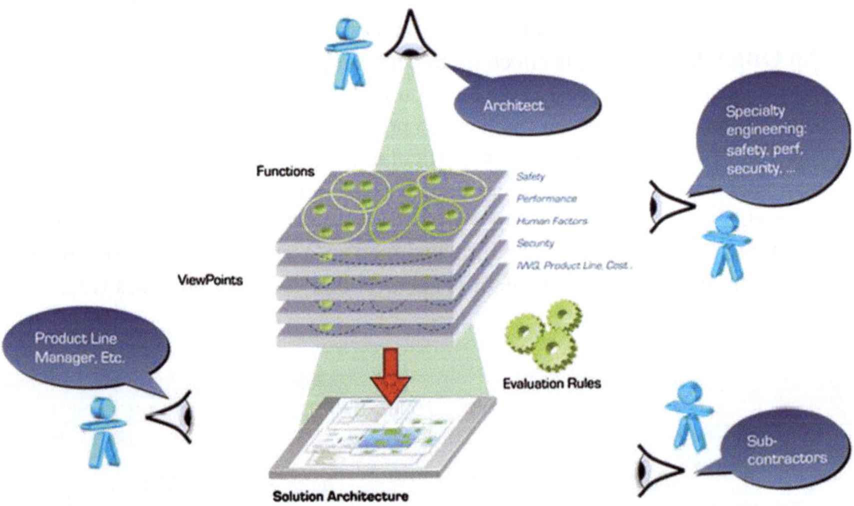

Fig. 11 Capella, the Thales "multi-points of view" System modeler

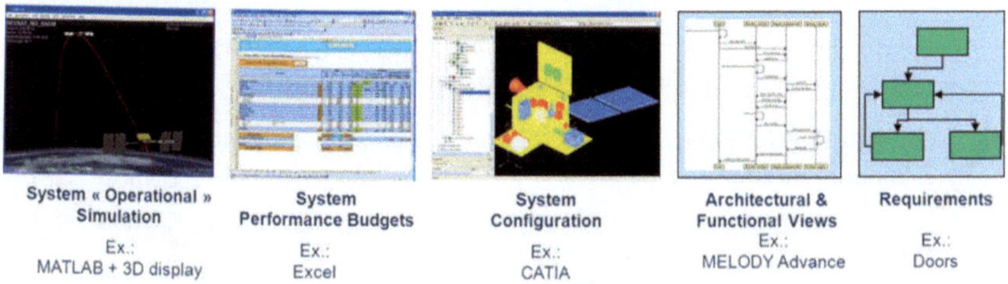

Fig. 12 Example of integrated engineering environment to support Co-Engineering

3 When and How to Practice Co-Engineering?

This chapter intends to illustrate through typical scenarios, how to assess the opportunity to implement the practice based on predefined "Co-engineering implementation criteria".

Note: Concerning facilitation, the involvement of a facilitator is mainly relevant for first experimentations of a given scenario (especially to enable exchange, mutual understanding and collaboration for an effective decision-making).

3.1 The Co-Engineering Implementation Criteria

These criteria intend to justify ("fitted to relevant") any Co-Engineering practice before its implementation; they must be reviewed and agreed by concerned managers whilst the strategy for system engineering (all activities included) is defined:

- **Stake**: "what is the objective?"; "what are the Business constraints to face to?", "what's the shared vision?"
- **Scope**: "What are the concerned Trough Life Cycle engineering activities"
- **Stakeholders** (the Co-Engineering team)—"Who must be involved" for acting and decision-making
- **Added-value:** "what is the observed value of the Co-Engineering practice?"
- **Level(s):** "what are the appropriate levels" so the resources (means, facilities,) to be available (and so invested beforehand)

3.2 Typical Scenarios for Co-Engineering Implementation

Figure 13 illustrates four fruitful approaches applied within Thales

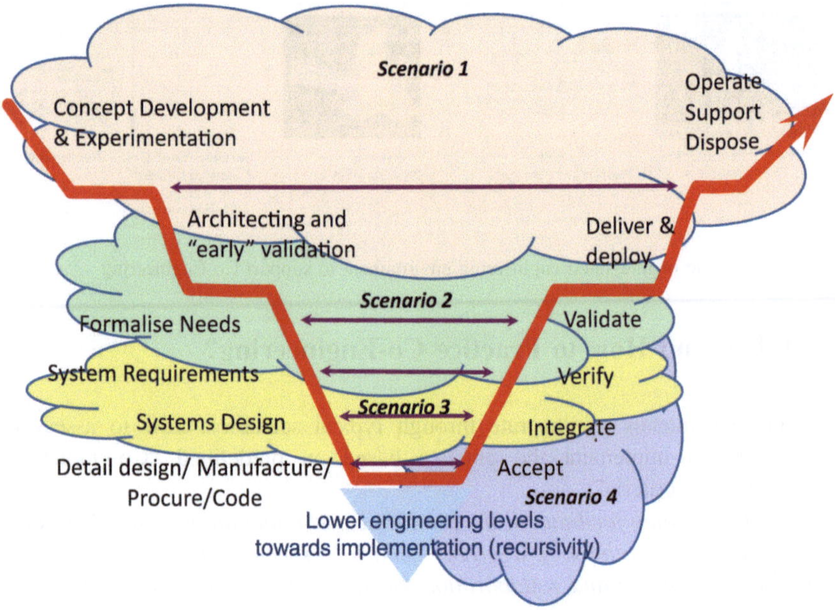

Fig. 13 Four fruitful Co-Engineering scenarios for system engineering

Scenario 1  "Consolidate the Operational fit for purpose"

- Stake: Reinforce the competitiveness, in Bid phase, by mastering the Operational fit for purpose. The contract includes Development, Deploy, Install, Support the "system". The offer must address the feasibility of a future transition to capability to "Operate" while ensuring that disposal will be in compliance with current regulations.
- Co-Engineering scope: "Architecting & Early validation" activities in coherence with "Deliver, Deploy, Operate, Support, Dispose" activities. The shared goal is to comply with Operational needs, usages and expected performances in an adaptive way (required capabilities can question the current Technical Offer, operational organisation being in full transformation).
- Co-Engineering team: Acquirer, Design authority, Operational expert/user, System/Support/IVVQ Engineering managers.
- Co-Engineering added value: Facilitated the capture (and mutual agreement on) of operational concepts (CONOPS, CONEMP, CONUSE) allowing to master the required capabilities and performances towards the expected global performance. Identified Architecture key drivers and highly critical non-functional constraints. Developed mind set of "Design for Operation, for Deployment, for Installation, for maintenance, for Disposal".

- Relevant levels: Level 4 recommended due to necessary operational simulations to be performed, based on dedicated tools.

Scenario 2  **"Early validate the certification"**

- Stake: Optimise the Certification activity while minimizing "time to delivery". Certification is a main constraint to transition into Service (Operate). Related Test campaigns are costly; optimisation is required.
- Co-Engineering scope: "Architecting & Early validation" activities, as source of references for "Formalize Needs" activity, in coherence with "Validate" activity. The shared goal is to comply with Certification constraints and reduce as far as possible Integration, Verification, Validation, Qualification and Certification [IVVQC] activities whilst providing with necessary evidences of conformity.
- Co-Engineering team (minimum): Acquirer, Design authority, Architect, Operational expert, Certification authority, IVVQ manager.
- Co-Engineering added value: Facilitated the capture (and the mutual agreement on) operational scenarios addressing the Certification and includes them in the contract. These inputs contributed to de-risking via "early validation" (based on simulation capabilities) the Certification process. Significant reduction of IVVQC activities has been observed. Acquirer's confidence increased.
- Relevant levels: Level 3 minimum; Level 4 may be useful to take benefit of powerful Engineering tools (domain simulations notably), as appropriate.

Scenario 3  **"Secure Integration and Verification"**

- Stake: Enforce the integrability and the verifiability of the system and implicitly secure Development delays and costs.
- Co-Engineering scope: "Systems Requirements" and "System Design" activities in coherence with "Integrate" and "Verify" activities. The shared goal is to establish while defining the system, a de-risked Integration and Verification strategy.
- Co-Engineering team: System Architect, IVVQ manager, Systems Engineering manager, necessary specialists (for non-functional criticalities).
- Co-Engineering added value: Secured the execution of "Integration & Verification" activities from the "System requirements" phase (notably when their validation) then during the "Systems Design" by considering integrability and verifiability as "Key Design" drivers ("Design for Testability" approach).
- Relevant levels: Level 2 is sufficient; (as appropriate) level 3 optimizes decision-making and level 4 enforces the decision via powerful Engineering tools.

Scenario 4  **"Optimize cross-disciplines IVV strategy"**

- Stake: Optimize whole IVV strategy so as to secure it at minimum cost; break the potential "silo" between engineering levels and increases synergy.
- Co-Engineering scope: "IVV" activities (multi-levels): a global vision for Systems IVV, Software and Hardware IV strategy (IVV Test campaigns, necessary resources, multi-levels synchronization).
- Co-Engineering team: System IVV manager, Software IV manager, Hardware IV manager, necessary specialists for critical specialties and certification constraints.
- Co-Engineering added value: Reinforced confidence and efficiency (avoiding recursive tasks on same scope) within the integrated IVV team. It facilitated the elaboration of a collaborative strategy notably the allocation of System tests campaigns to the relevant level (e.g. System [HMI]—"Agile development" to Software IV team, being responsible of).
- Relevant levels: Level 2 is sufficient; level 3, as appropriate, notably in case of decision-making in an initially conflicting context.
  Note: This scenario is particularly efficient in case of "Product Policy" context. By integrating the (Domain) Product IVV manager in the Co-Engineering team, it optimizes IVV tasks—for the generic and reusable features and a given variability—within teams concerned by:
- Engineering of "Product **for** Projects" (Domain investment, marketing and competiveness value).
- Engineering of "Product **of** Projects" (Business and Project consequent value).

4 Conclusion

The benefit of Co-Engineering practice within Thales is indisputable, as illustrated by previous scenarios. In a few words, it contributed to:

- Increase the **efficiency** and the **maturity** of engineering teams, securing lead time and costs
- Improve the **satisfaction** and the **confidence** of the customer:
  - by enforcing the **Quality** of engineering deliverables,
  - by optimizing the product **Life cycle Cost**, that is a key discriminating factor for both "**Win Bid**" and "**keep one's Customer**"
- Stimulate the "**value pulled**" approach

- And implicitly:
  - reinforce **competitiveness**
  - attract and keep "**talents**"

Nevertheless, the adoption and the massive deployment of this practice have to face to **basics breaks**: as example, investment costs (when an effective ROI?), real conviction (global value vs. own value?), human behaviour (Is this "shared" vision really mine?).

To mitigate these breaks, Thales implemented a **dedicated support** aiming to facilitate and reinforce the Co-Engineering deployment.

At Thales Group level, thanks to:

- **Sponsoring** (Head of Engineering communication to Engineering stakeholders by valorisation of Co-Engineering as main lever for breaking the « silo » effect)
- **Synergy with Lean Engineering** deployment (notably on human behaviour and facilities aspects)
- **Dedicated Training** within "Thales Université": The training course has been designed for improving skills in Co-Engineering for organisation & animation, addressing:
  - Collaborative Engineering across project phases (Bid and Project)
  - Methods, Types of meetings
  - Facilities and Infrastructure enhancing Co-Engineering approach
  - (e.g.: CDF Room, Creativity centre, OBEYA Room)
  *25 sessions of this course have been delivered in **2014** to train about **300** engineers, and the same volume is planned in 2015*
- **Dedicated events** (called "Co-Engineering" days)
- **Capitalisation** ("Good practices" and "pitfalls to avoid" based on RETEX, community of Interest)

At "Entity" level:
Equivalent approach has been applied at "local" level [sponsoring, capitalisation, events, mutualisation in necessary means and facilities, participation to Thales Université training (as trainee and/or as co-designer, testimony provider)].

As example of testimony capture, **the "TSA Co-Engineering day" (from Edmond Tonnellier, organizer)**

> Thales Systems Aéroportés (TSA) has organized an Engineering Day concerning the Co-Engineering to exchange experiences and points of view on Co-engineering with representatives stemming from various jobs (businesses, engineering system, software, material, specialties, production, support) with attendees coming from various sites of DMS (Division Missions Systems, Thales Avionics, the Corporate, Global Thales Services and Thales Université).
>
> Indeed, to meet constraints of deadline, teams involved in a project lead their works in parallel, rather than in a sequential way.

Co-engineering has for ambition to improve the degree of collaboration between teams. It encourages in particular a shared vision of the project, with common objectives clearly definite. This shared vision covers the entirety of the life cycle of the system, the design in the deployment, up to the operational support, and involves all the stakeholders of the system, the customer to the suppliers, the purchases in the production and the service.

The human dimension was advanced, as one of the first success factors in the implementation of an approach of Co-engineering.

Today, Co-engineering is a necessity for all the new large-scale projects of Thales. There is an imperative of deployment, in particular on the international projects, requiring the implication of teams of engineering geographically taken away. Co-engineering increases in importance.

References

1. The ESA Concurrent Design Facility Concurrent Engineering Applied to space mission assessments, CDF, ESA/ESTEC, Noordwijk, NL, CDF Info Pack (2015)
2. Santos, P.I.N., Raposo, A.B., Gattass, M.: A software architecture for an engineering collaborative problem solving environment. Software engineering workshop, 2008 (SEW '08), 32nd Annual IEEE
3. Park, J.-P., Yang, S.-W., Kwon, K.-E., Choi, Y.: Collaborative engineering and product quality assurance based on integrated engineering information management. Smart Manufacturing Application, 2008 (ICSMA 2008)
4. Wang, C.-B., Chen, Y.-M., Chen, Y.-Z.: A distributed knowledge model for collaborative engineering knowledge management in allied concurrent engineering. In: Engineering Management Conference, 2002 (IEMC '02)
5. Sriram, P.K., Alfnes, E., Kristoffersen, S.: Collaborative engineering: a framework for engineering-to-order companies. In: Collaboration Technologies and Systems (CTS), 2014
6. Martin, E., Fabrice, M.N.: Conceptual modeling and generator framework for multidisciplinary and collaborative product lifecycle management. In: Computer Supported Cooperative Work in Design, 2009 (CSCWD 2009)
7. Mosher, T.J., Kwong, J.: The Space Systems Analysis Laboratory: Utah State University's new concurrent engineering facility. In: Aerospace Conference, 2004
8. Landauer, C., Bellman, K.L.: Collaborative system engineering and integration environments. In: Enabling Technologies: Infrastructure for Collaborative Enterprises (1996)
9. McQuay, W.K.: Distributed collaborative environments for systems engineering. In: Digital Avionics Systems Conference, 2004 (DASC 04)
10. Karvonen, I., Uoti, M., Granholm, G.: Application of systems engineering in a collaborative environment. In: Engineering, Technology and Innovation (ICE), 2012
11. Del Rosario, R., Davis, J.M., Keys, L.K.: Concurrent and collaborative engineering implementation in an R&D organisation. In: Engineering Management Conference, 2003 (IEMC '03)
12. McQuay, W.K.: A collaborative engineering environment for 21st century avionics. In: Aerospace Conference (1998)
13. Bechina, A., Brinkshulte, U.: Towards a distributed collaborative product engineering. In: Industrial Technology, 2003 IEEE International Conference
14. Beebe, B.W., Shedden, J.S.: A collaborative application of systems engineering. In: Aerospace and Electronics Conference (NAECON), 2009
15. Lu, S.C.-Y., Elmaraghy, W., Schuh, G., Wilhelm, R.: A scientific foundation of collaborative engineering. In: CIRP Annals—Manufacturing Technology, 2007

16. Lu, S.C.-Y., Cai, J., Burkett, W., Udwadia, F.: A Methodology for collaborative design process and conflict analysis. In: CIRP Annals—Manufacturing Technology, 2000
17. Willaert, S.S.A., de Graaf, R., Minderhoudc, S.: Collaborative engineering: a case study of Concurrent Engineering in a wider context. J. Eng. Technol. Manage. **15**, 87–109 (1998)
18. [DoD IPPD], Department of Defence Integrated Product and Process Development Handbook, August 1998
19. [INCOSE SE HB], Systems Engineering Handbook, a guide for system life cycle processes and activities, V3.2.2, International Council on Systems Engineering (INCOSE), INCOSE-TP-2003-002-03.2, Oct 2011
20. [CMMI], CMMI® (Capability Maturity Model® Integration) for Development, V1.3, November 2010

Simplification Principles in the Design of Cyber-Physical System-of-Systems

Hermann Kopetz

Abstract Systems-of-systems are built by the integration of autonomous existing systems, called constituent systems (CS), in order to provide new synergistic services and improved economic processes. When integrating cyber-physical systems (CPSs), the interactions among the constituent systems are not confined to the exchange of messages in cyber space but are also realized by a stigmergic information flow in the physical world. The size of the CPSs and the multitude of the interactions among the CPSs lead to an enormous cognitive complexity of the behavior of the CP-SoS and make it difficult to reason about the correct operation of a cyber-physical system of systems (CP-SoS). It is the objective of this paper to present some simplification principles that help to reduce this cognitive complexity of a CP-SoS.

1 Introduction

The domain of Systems-of-Systems (SoS) is a relatively new field of computer science that is concerned with the cross-organizational design, integration and operation of large information processing systems that are composed of heterogeneous existing or new autonomous constituent systems (CS) [1]. In many cases a CS is a cyber-physical system (CPS), i.e., it consists of interacting computer systems, physical machines and humans. It is assumed that the widespread integration of existing CPSs and data bases that bring about cyber-physical systems-of-systems (CP-SoSs) will make better use of available information, lead to new insights, improve current economic processes and thus provide new synergistic services to create greater wealth.

We take the definition of an SoS from the work of Jamshidi [2]: *System of Systems are large-scale integrated systems that are heterogeneous and indepen-*

H. Kopetz (✉)
Institut Für Technische Informatik, TU Wien, Wien, Austria
e-mail: H.Kopetz@gmail.com

G. Auvray et al. (eds.), *Complex Systems Design & Management*,
DOI 10.1007/978-3-319-26109-6_3

dently operable on their own, but are networked together for a common goal. The cyber space of a CP-SoS is formed by the interacting computer systems that are part of the constituent systems (CS), while the controlled equipment and the human users/operators/managers form the physical constituents of the CP-SoS. Table 1 characterizes a CP-SoS by listing some distinguishing properties of a CP-SoS compared to those of a classic monolithic CPS [3, 4]. If we look at this table we see that the listed characteristics of a CP-SoS violate many of the fundamental assumptions that are taken for granted in the established system design process. For example, there is no fixed specification, coordinated evolution, or final acceptance test of a CP-SoS.

It is the objective of this paper to present some simplification principles in order that the cognitive effort—and thus the elapsed time—needed to understand the behavior of a CP-SoS can be reduced. The adherence to these simplification principles should result in a more intelligible design.

We base our work on the assumption that an explanation of the behavior at the CP-SoS level must be based solely on the observation of the relied-upon interface behavior of the constituent CPSs that is specified in the service level agreements (SLA). It is not advisable to refer to knowledge about the internals of a CPS, since the internals of a CPS may be changed without changing the relied-upon interface behavior.

This paper starts by elaborating on the concept of *simplicity* that we consider the antonym of *cognitive complexity*. The introduction of a global notion of time that is used as a control variable reduces the complexity of a design. We then look at the properties of the relied-upon-interfaces of a CPS that form the backbone of the CP-SoS architecture. The CSs of a CP-SoS interact by *message-based information items* in cyberspace and by stigmer*gic information items* transported in physical space. The proper handling of the interface-state, the topic of Section four, can decouple future behavior from past behavior and thus contribute to a simplification of the reasoning about the behavior of a CP-SoS. Section five is devoted to cope with the observation that *faults are normal* in a large CP-SoS. The final Section six summarizes the contents of this paper by listing a number of concrete simplification principles for the design of a CP-SoS.

Table 1 Comparison of CPS and CP-SoS

Characteristic	CPS	CP-SoS
Scope of system	Fixed (known)	Not known
Requirements and spec.	Fixed	Changing
Evolution	Version control	Uncoordinated
Testing	Test phases	Continuous
Faults (physical, design)	Exceptional	Normal
Architectural style	Single	Multiple
Governance	Central	Uncoordinated
Emergence	Insignificant	Important

2 Simplicity

The purpose of applying a simplification strategy to a design of a system is to make the evolving system *more intelligible*, i.e., the *cognitive effort*—and consequently the elapsed time—needed to *understand* the behavior of the system should be reduced. *Simplicity* is thus of utmost economic significance, since the amount of time—the *dominant cost factor*—needed to design, use, maintain and change a system is cut down.

Simplicity is a *relation* between a scenario and an observer and not a *property* of a scenario [5]. A novice to a field can consider the behavior of a scenario *complex* even if an expert judges the same behavior as *simple*. Simplicity is the antonym of *cognitive complexity*. The refinements of the notion of complexity, depicted in Fig. 1, are not fully orthogonal. A scenario that has a high *dynamic complexity* will also have an elevated *cognitive complexity*.

What does it mean that *an observer understands the behavior of a system*? In his research on *the nature of understanding* Craik [6] posits that an observer has confidence that he *understands* a system as soon as he has developed a *mental model* commensurate with his cognitive capabilities that allows him to reason about the *cause-effect relationships* of the observable events at the interfaces of the system. *Causal reasoning* is a fundamental method of operation of the human mind in order to survive in a dynamic environment. How can we support the conception of such a mental model that supports causal reasoning?

Causal order presupposes *temporal order*. *Event A* can only be the cause of *event B* if *event A* happened temporally before *event B*. Causal event analysis is simplified if all events are timestamped with a global time. Based on these timestamps it is possible to eliminate all those events of the set of events that cannot be causally related. The time-stamping of all events with a global time of adequate

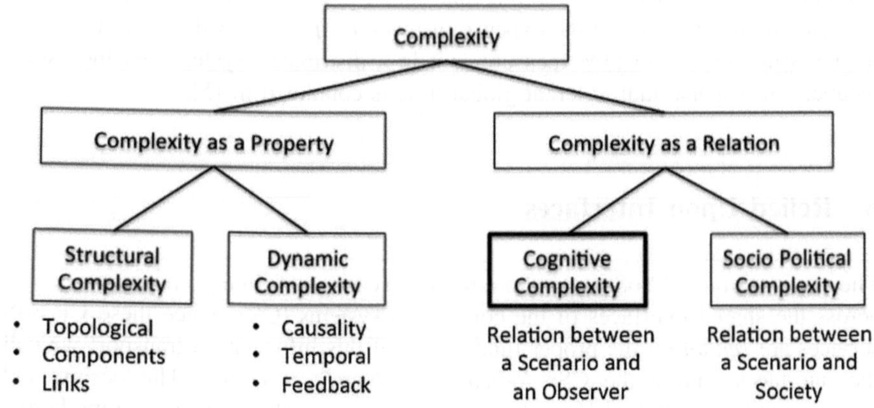

Fig. 1 Refinement of the notion of complexity

granularity can simplify the understanding of the behavior of a large system-of-system, as the following example demonstrates.

On August 14, 2003 a major power blackout occurred in parts of the US and Canada. In the final report [7] about this blackout it is stated: *A valuable lesson from the August 14 blackout is the importance of having time-synchronized system data recorders. The Task Force's investigators labored over thousand of data items to determine the sequence of events, much like putting together small pieces of a very large puzzle. That process would have been significantly faster and easier if there had been wider use of synchronized data recording devices.*

A global time among the CPSs of a CP-SoS is also needed for the implementation of time-triggered communication protocols (TTCP) among the CPSs. The understanding of the behavior of TTCPs is *significantly simpler* than the comprehension of the behavior of an event-triggered communication protocol, because the instants for sending a message in a TTCP depend solely on the progression of the global time and are *independent* of any *change of value* (an event) in the data domain. This complete *decoupling of the temporal behavior* of the communication system from the *behavior in the data domain* that is not realizable if event-triggered communication protocols are in use, easies significantly the analysis of the temporal behavior of time-triggered systems.

Real-time data, e.g., the state of a *traffic light*, is invalidated by the progression of time. Using a real-time data element outside its *temporal validity interval* may be the cause for an accident. It is therefore good practice to provide a global timestamp that denotes the *termination of the temporal validity interval* as part of every real-time information item.

The operation of the clocks in the nodes of a distributed CPS can never be fully aligned. Due to the unavoidable *digitalization* and *synchronization errors* it is possible that the *true order of events* that occur in close temporal proximity to each other in the physical environment is not reflected in the recorded timestamps generated by the use of synchronized clocks. An improved precision of the clock synchronization will alleviate these effects, but will never fully resolve this dilemma. The introduction of a *sparse time* base will establish the *consistency of the time-stamps* in a CP-SoS at the expense of *temporal fidelity*. A detailed discussion of the issues related to time measurement in a distributed system and the establishment of a sparse fault-tolerant global time is contained in [5].

3 Relied-Upon Interfaces

The services of a CP-SoS come about by the exchange of timely information items across the shared interfaces of the constituent systems (CS). Since these CP-SoS services are *reliant* on the proper functioning of this information transport we call the interfaces among the CSs *Relied-Upon Interfaces (RUI)*. The design and placement of these *Relied-Upon Interfaces* has a huge effect on the comprehensibility of the behavior of the CP-SoS.

3.1 Information Versus Data

According to Ref. [8] *An interface is a shared boundary across which two components of a computer system exchange information.* The information that is transported across an interface consists of one or more *information items* or *Itoms.*

The *semantic content* of (or the *information* contained in) an *Itom* reports about *a proposition relating to some entities in the world* [9]. An Itom consists of *data* and an *explanation of the data.* In cyber space *data* is represented by a *bit-pattern* that can be generated by some data acquisition process, e.g., by a *sensor*. In this case, the *design of the sensor* determines how the acquired bit pattern has to be interpreted, i.e., provides for the *explanation* of the data.

The data representation of the *semantic content* of an Itom depends on the *context*. For example, in the US temperature is represented by *degrees Fahrenheit,* while in Europe temperature is represented by *degrees Celsius*. In a CP-SoS the involved CSs can be operating in differing contexts, e.g., in the *US* and *Europe*. As a consequence, the *same semantic content* (*information*) can be represented by *different bit-patterns* (data) at the two sides of the interface, causing a *property mismatch*. Such a *property mismatch* has been a cause of severe accidents.

Since an *Itom* is a higher-level concept than the *data* in an Itom, we propose to use *Itoms* in the specification of the RUIs among the CSs. According to Kopetz [9] the full specification of an Itom has to provide answers to the following questions:

- **Identification**: *What entity is involved?* The entity must be clearly identified in the space-time reference frame.
- **Purpose**: *Why is the data created?* This answer establishes the link between the raw data, the refined data and the purpose of the CP-SoS.
- **Meaning**: *How has the data to be interpreted by a human or manipulated by a machine?* If the answer to this question is directed towards a human, then the presentation of the answer must use symbols and refer to concepts that are familiar to the human. If a computer acquires data, then the explanation must specify how the data must be manipulated and stored by the computer.
- **Time**: *What are the temporal properties of the data?* Real-time data must include the instant of observation in the entity. In control applications it is helpful to include a second timestamp, a *validity instant* that delimits the validity of the control data as part of the Itom.

3.2 Interface Types

In a CP-SoS the CSs can exchange information across interfaces to two different types of channels, *message based channels*—we call them *Relied upon Message Interfaces* (*RUMI*)—and *stigmergic channels*—we call them *Relied upon Physical Interfaces* (*RUPI*) [10].

Table 2 Stigmergic versus message-base information flow [8]

Characteristic	RUPI (stigmergic)	RUMI (cyber message)
Information type	Properties of entities	No restriction
Inform. transfer	Pull	Push
Tense	Present	Past, present, future
Observation mode	Direct	Indirect
Observation delay	None	Existent
Comm. delay	Unbounded	Bounded
Source	Unknown	Known
E-dynamics	Considered	Not considered
Representation	Single context	Multiple contexts

A *stigmergic channel* is present if one CS acts on the physical environment and changes the state of the environment and later on another CS observes the changed state in the environment. Consider, for example, the coordination of cars on a busy highway to realize a smooth flow of traffic. In addition to the direct communication by signals among the drivers of the cars (e.g., the blinker or horn), the *stigmergic information flow* based on the observation of the movement of the vehicles on the road (caused by the actions of other drivers) is a primary source of information for the assessment of a traffic scenario. Table 2 compares the characteristic of stigmergic channels with those of message-based channels. An important characteristic of stigmergic information is the consideration of up to date *environmental dynamics*, i.e., processes in the environment that change the value of an information itom.

The biologist Grasse introduced the term *stigmergy* to describe the indirect information flow among the members of a termite colony when the nest building activities are coordinate. Grasse [11] showed that the coordination of the termites is not achieved by direct communication among the workers, but by indirect communication based on stimuli from the emerging physical nest structure in the environment.

The identification of the Itoms that are transported via stigmergic channels of an SoS is important, because these Itoms often form the *missing link* in a feedback or feed-forward control loop. Unidentified control loops can be the reason for disturbing emergent behavior.

3.3 Interface Placement

The placement of the Relied Upon Interfaces (RUI) requires special attention:

- Minimize the dependence of the CSs on each other and provide maximum autonomy to the CSs.
- Hard real-time requirements should be serviced within a CS.
- Sensor specific data should be preprocessed within a CS to produce timestamped data in a standardized form at the RUI.

In order to increase the stability of RUIs, the RUIs should be placed at the boundaries of subsystems that can absorb foreseeable changes in the environment since changes are expected to occur in an evolving environment.

Take, for example, a sensor system that is designed to deliver the value of a specific physical quantity. A change of the sensor system to another sensing method that delivers the same physical quantity should not have any effect on the corresponding RUPI.

3.4 Interface Model

The reasoning about the behavior of an SoS has to be based solely on the specification of the behavior at the relied upon interfaces (RUI) and *must not* require an analysis of the internals of the interfacing CSs. For this purpose a *fully specified interface model* must be contained in the *Service Level Agreement* (SLA) that describes the meaning and constraints of the Itoms that pass the interface and explains the services provided by the interfacing CS [12]. In addition to the syntactic specification of the data elements, a sound interface model must provide a complete explanation of the acceptable data domains, the meaning of the data and their temporal properties, such as the permitted *temporal validity interval* of real-time control data, from the point of view of an interface user. A functional description of the provided data transformation, based on observable data, must also be part of the interface model.

A *service user* should base its work only on the information contained in the interface specification of the SLA and should not make any assumptions about the internal operation of a CS. As long as a change in the internal operation of a CS does not affect the RUI it is of no concern to a service user. This property is important from the point of view of *independent evolution* of a CS.

Many interface specifications are deficient of a precise specification of the temporal properties of the RUI. A modification within a CS that does not change the function of the services but modifies the (unspecified but nevertheless assumed) temporal properties can then give rise to unintelligible malfunctions at the SoS level.

The precise interface model of the RUI, contained in the SLA, is essential for the detection of CS-errors (see Sect. 5.2).

4 State Management

The proper management of the *state information* referred to in the interface model is of particular concern for the understanding the behavior of a CP-SoS, since a clear notion of state *segments* the time-line into *past behavior* and *future behavior* and thus simplifies the causal reasoning about the behavior of an SoS.

4.1 *Definition of* State

We take the definition of the *state* from [13]:

> The state enables the determination of a future output solely on the basis of the future input and the state the system is in. In other word, the state enables a "decoupling" of the past from the present and future. The state embodies all past history of the given system. Apparently, for this role to be meaningful, the notion of the past and future must be relevant for the system considered.

Taking this definition, the concept of *state* is only meaningful if a precise notion of time is available in the CP-SoS, since the state is a function of time and changes as time progresses.

Many cyber-physical systems exhibit a *periodic* behavior, e.g., repeatedly traversing a control loop. Finding a periodically recurring *reintegration instant* in a control loop where the state data is minimal is an important design consideration for the effective storage and recovery of state data.

4.2 *Stateless Versus Statefull Services*

We call a service-session that is executed by a *service-providing CS* across a RUI *state-agnostic* if the *input data* and the *required state data* is taken as input and the *output data* and the *updated state data* is provided as output of the service, given that the time-stamp of the termination of a *session is temporally before* the timestamp of the start of the next session.

In the time-interval "*termination of a session, start of the next session*" the *service-providing CS* is then *stateless*. Otherwise, the service-session is called *statefull.*

From the point of view of cognitive complexity, *state-agnostic service sessions* are *simpler* to understand than statefull service sessions because no assumptions have to be made about the value of some unobservable state variable hidden inside a CS. It is therefore a good design practice to increase the size of the observable input/output data of a service by the necessary state data in order to be able to realize a *state-agnostic service session.*

The updated state data can be sent periodically to an *independent monitoring system* for storage. The monitoring system must analyze the state and perform a continuous update of the state to bring it into agreement with the state of the evolving physical environment, observed by a sensor of the monitoring system (see Sect. 5.2). A CS that is recovering after the occurrence of a transient failure can then acquire the current state data from the monitoring system.

5 Faults Are Normal

Looking at Table 1, the characteristic that we consider most relevant for the design of a CP-SoS is the statement *Faults are normal in a CP-SoS*, since this characteristic influences significantly the outlay of the structure of an SoS. In any large system, like a CP-SoS, the occurrence of transient or permanent hardware faults and undetected design errors is the norm, rather than the exception.

It is crucial to differentiate between *easily* reproducible software errors (*Bohrbugs*) and *difficult* to reproduce software errors (*Heisenbugs*) [5]. From the point of view of *fault occurrence*, transient faults, i.e. *transient hardware faults* or *Heisenbugs*, are most probable. Transient faults do not damage the hardware permanently, but corrupt the *state*. If mechanisms are provided in the design such that *corrupted state* can be *detected and repaired quickly and autonomously*, the consequences of the occurrence of a transient fault can be mitigated without any external repair action.

5.1 Fault-Containment and Error Detection

The first steps in the design of a fault-tolerant system concern the establishment of proper *fault-containment units* (FCU) and *the identification of the considered faults of the FCUs*. The physical and functional design must ensure that a single fault will only affect a single FCU. In a fault-tolerant CP-SoS every constituent system should be an FCU. The most likely failure mode of an FCU is *fail silence*—the FCU stops to function. Fail-silence failures can be detected *solely* in the temporal domain.

A fault in a CS can propagate to another CS by the transport of erroneous information items (Itoms) across an interface. In cyber-space, Itoms are transported by messages. In order to detect an erroneous message, we need a second FCU that is independent of the FCU where the fault occurred.

A message can be erroneous in the *time domain* or in the *value domain*. Message errors in the time domain—the omission of an expected message or the untimely production of a message—can be detected by placing a time-triggered communication system, such TT-Ethernet [14], between the CSs of a CP-SoS (Fig. 2). A time-triggered communication system acts an independent FCU that contains information about the correct temporal performance of the interfacing CSs. If a CS violates its temporal specification the time-triggered communication system can detect the error immediately.

The detection of errors in the value domain requires redundant *application specific information*. Application specific information can be contained in an *independent monitoring system* (*IMS*) that is familiar with the *Service Level Agreements* (*SLA*) of the constituent systems (Fig. 2). The SLA should specify the

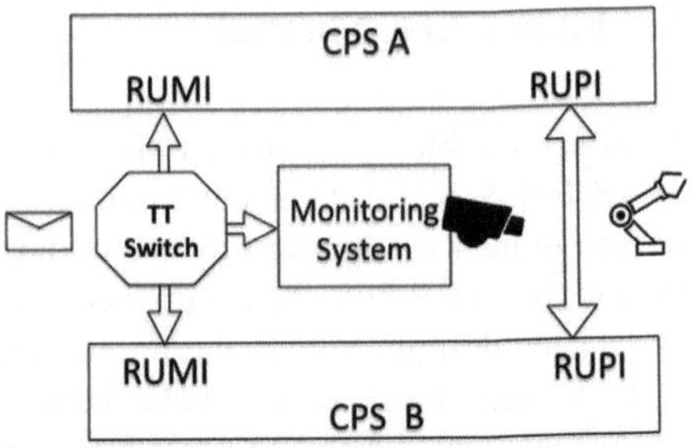

Fig. 2 Independent monitoring system

value domains of the Itoms that cross the RUMIs and should contain an executable version of the interface model (see Sect. 3.4). Based on this information the IMS can perform a gross value-check of the messages that cross the RUMI and can detect blatant value violations.

5.2 *Independent Monitoring System (IMS)*

It is good practice to monitor and document all information flows across the interfaces of a CS non-intrusively by an independent monitoring system.

Appropriate independent sensors of the IMS can observe the timely information flow across the RUPI that is part of the interface model (Fig. 2). For example, if the RUPI deals with the actions of a machine (e.g., a robot), a camera can capture the actions in the physical environment and perform image processing to find out whether the intended actions are happening in the physical environment at the intended instant. If the images are time-stamped with the global time, than the actions in the physical world can be aligned with the time-stamped message exchanges in cyber-space in order to gain a better understanding of the behavior of the whole CP-SoS as time unfolds.

In order to improve the error-detection capability of the IMS the IMS should also contain *models of the actuators* at the *physical interfaces* that inform about the time-delays between an actuator-command, originating in cyber space, and the consequent physical action, observable by the IMS via its sensor system.

An independent monitoring system can simplify *the causal analysis* of the behavior of a CP-SoS for the following reasons:

- The *nonintrusive documentation* of all input and output message and physical actions of the interfacing CS, including their timestamps, forms the basis for the later reconstruction and analysis of the *input/output behavior* of the interfacing CS.
- Any violation of the SLA by a CS, either in cyber space or in the physical space, is immediately detected and documented for future analysis.

- The IMS can store the current state of a CS and provide this state data, modified by the current information captured from the observation of the physical environment, to a recovering CS.
- It is possible to *replay* the documented input/output history of a CS to reproduce an observed anomaly.

6 Simplification Principles

A number of principles for the simplification of embedded computer systems have been published in the literature [5]. In addition to these general principles we summarize our analysis of the cognitive complexity of CP-SoSs by providing a list of specific simplification principles for the design of a CP-SoS:

- **Global Time**: The time-stamping of all observable events at the interfaces of the CSs with a CP-SoS global time simplifies the reasoning about the causal order of events, helps to get an understanding of the behavior of the CP-SoS and supports an immediate detection of a violation of the temporal properties of the service level agreement (SLA) by a CS.
- **Relied upon Interface (RUI)**: The design and placement of the *relied-upon interfaces RUI* that form the backbone of a CP-SoS architecture should be contrived to maximize the autonomy and independence of the CSs and to minimize necessary modifications in case of changes in the external environment. Relied-upon interfaces should be placed at the boundaries of subsystems that can absorb foreseeable changes in the environment.
- **Time-triggered Communication Protocols**: The use of time-triggered protocols for the data transport among the CPSs eliminates the dependence of the temporal behavior of the communication system from the behavior in the data domain.
- **Information versus Data**: Using *information items* (*Itoms*) instead of *data items* to specify the information transport across the relied-upon interfaces (RUI) of a CP-SoS puts the description at the higher level of abstraction that simplifies understanding.
- **Message-based versus Stigmergic Interfaces**: The identification of all message-based itoms that are transported across the relied-upon message interfaces (RUMI) and all stigmergic information itoms that are transported across the relied-upon physical interfaces (RUPI) helps to detect hidden feedback loops.
- **Interface Model**: The interface model contained in the Service Level Agreement (SLA) must establish the meaning, the functional relationships, and the temporal properties of the input/output Itoms that cross the interfaces of a CS.
- **Identification of State**: The identification of the *state* at planned reintegration instants segments past behavior from future behavior and simplifies the reasoning about the behavior. Finding a periodically recurring *reintegration instant*

in a control loop where the state data is minimal is an important design consideration for the effective storage and recovery of state data.
- **State-agnostic Service Sessions**: From the point of view of cognitive complexity, *state-agnostic service sessions* are *simpler* to understand than statefull service sessions.
- **Fault Containment**: In a CP-SoS every constituent system should be designed to be a fault-containment unit (FCU).
- **Error Detection**: The propagation of errors across RUIs must be hindered. The most likely failure mode of an FCU is *fail silence*—the FCU stops to function. Fail-silence failures can be detected *solely* in the temporal domain.
- **Time-triggered RUMI**: A *time-triggered message switch* that is placed in the connections between RUMIs can detect errors in the temporal domain and can route the interface *Itoms* to an *independent monitoring system* for the detection of errors in the value domain.
- **Independent Monitoring System (IMS)**: It is good practice to monitor and document all information flows across the RUMIs of a CS non-intrusively by an IMS and to equip the IMS with sensors to observe the information flow across the RUPIs.

7 Conclusion

The main struggle in the design, operation and maintenance of a large Cyber-physical Systems-of-Systems (CPSoS) is related to the *cognitive effort* (measured in elapsed time) required to understand the behavior of the CP-SoS under normal and fault conditions. In this paper we have discussed a number of principles for the design of an SoS that should help to reduce this cognitive effort. We propose to invest in additional hardware resources, such as a time-triggered switch between the message interfaces of constituent systems and an independent monitoring system with sensors to observe the physical environment, in order to improve the error-detection capability of the system and to make the interface behavior of the constituent systems visible and reproducible. We feel that in most cases the cost of these additional hardware resources is small compared to the savings in engineering time that results from the additional information produced by these resources.

Acknowledgments This work has been supported, in part, by the European FP7 research project AMADEOS Grant Agreement 610535 on *Systems of Systems*. Many discussions with members of the AMADEOS Project are warmly acknowledged.

References

1. Gordod, A., et al.: System-of-system engineering management: a review of modern history and a path forward. IEEE Syst. J. **2**(4), 484–499 (2008)
2. Jamshidi, M.O.: Systems-of-Systems Engineering. Wiley, New York (2009)
3. Garro, A., Tundis.: System reliability analysis of systems and SoS: the RAMSA method and related extensions. IEEE Syst. J. **9**(1), 232–241 (2015)
4. Kopetz, H.: Systems-of-systems complexity. In Proceedings of the 1st Workshop on Advances in Systems of Systems, AiSoS 2013, EPTCS 133, pp. 35–39
5. Kopetz, H.: Real-Time Systems—Design Principles for Distributed Embedded Applications. Springer, Berlin (2011)
6. Craik, K.: The Nature of Explanation. Cambridge University Press, Cambridge (1967)
7. USC04.: US-Canada Power Outage Blackout Report. https://reports.energy.gov
8. Wikipedia on *Interfaces*
9. Kopetz, H.: A conceptual model for the information transfer in systems-of-systems. In Proceedings of ISORC 2014, pp. 17–24. IEEE Press, New York
10. Kopetz, H., et al.: Direct versus stigmergic information flow in systems-of-systems. Accepted for Publication in Proceedings of 10th Annual System-of-Systems Engineering Conference, SOSE 2105. IEEE Press, New York
11. Grasse, P.P.: La reconstruction du nid et les coordinations interindiviuelles chez Bellicositermes natalensis et Cubitermes sp. La theorie de la stigmergie. Insectes Sociaux. **6**, 41–83 (1959)
12. Erl, T.: SOA Principles of Service Design. Prentice Hall, Englewood Cliffs (2008)
13. Mesarovic, M.D.: Abstract system theory. In Lecture Notes on Control and Information Science, vol. 116. Springer, Berlin (1989)
14. SAE *Standard AS6802 TT Ethernet*. http://standards.sae.org/as6802

System Readiness Assessment (SRA) a Vade Mecum

Marc F. Austin and Donald M. York

Abstract As the complexity of systems increases, it is critical to develop a more comprehensive understanding of the development status, or readiness, of the system to aid more informed system-level technical and management decisions throughout the life cycle. Lack of comprehensive system thinking at the onset and failure at the integration points are two of the primary causes for unsuccessful system development. To measure system readiness, a greater emphasis must be placed on integration. This paper provides a vade mecum or handbook for System Readiness Assessment (SRA). It provides system-level metrics that give visibility over the development life cycle into the entire system and its interfaces. The SRA assessment criteria are described and an example is provided. The intended users include Program Managers, Systems Engineers, Independent Review Teams, and developers. The goal is to provide a fundamental understanding of how to conduct an SRA, as well as assist experienced users in maximizing the benefits of SRAs.

1 Introduction

This paper invites the reader to walk with us as we describe in a handbook fashion the details of a SRA process. SRAs help to improve performance management for systems and aid decision makers in identifying programmatic and technical risks. We describe the criteria and metrics used in an assessment, provide sample calculations, and explain the application of the System Readiness Level (SRL). This Handbook gives guidelines for effective implementation and use of the SRA process. Intended users include Program Managers, Systems Engineers, Independent

M.F. Austin (✉)
US DOD, The Pentagon, VA, USA
e-mail: mmfaustin@gmail.com

D.M. York
TASC, An Engility Company, Annapolis Junction, MD 02701, USA
e-mail: donald.york@engilitycorp.com

G. Auvray et al. (eds.), *Complex Systems Design & Management*,
DOI 10.1007/978-3-319-26109-6_4

Review Teams (IRTs), and developers. As the reader walks with us, we provide a fundamental guidance on conducting SRAs, as well as assist experienced users to maximize the benefits derived from performing SRAs.

As systems increase in complexity with a growing number of interfaces and integration and sustainment issues, it is critical to develop a comprehensive view of the system development status, or "system readiness." This will aid informed system-level technical and managerial decisions throughout the life cycle, reducing both programmatic and technical risk. To develop potential system-level metrics, a greater emphasis must be put on the integration between and among individual components since integration issues are one of the leading causes of system failure. During large-scale developments, it is critical to measure system readiness at multiple points along the life cycle to avoid the pitfalls when readiness is only assessed once or twice. SRAs are an innovative methodology that provides the system metrics that address the dramatically changing deployment and operational environment of modern day systems. The SRA process provides visibility over the entire development life cycle into the entire system and the system's interfaces as well as external entities.

The SRA process gives decision-makers awareness of a system's holistic state of maturity and quantifies the level of integration of a specific component with other components during system development. It helps the Program Manager determine the system's ability to produce its intended capability. In effect, it's like a System Development Mall and the SRA is like the mall map which tells you that "You are here!" (See Fig. 1) The assessment is a critical part of improving system performance management and reducing risk. Both recent academic efforts and internal expertise have been leveraged to develop this SRA process. Much of the

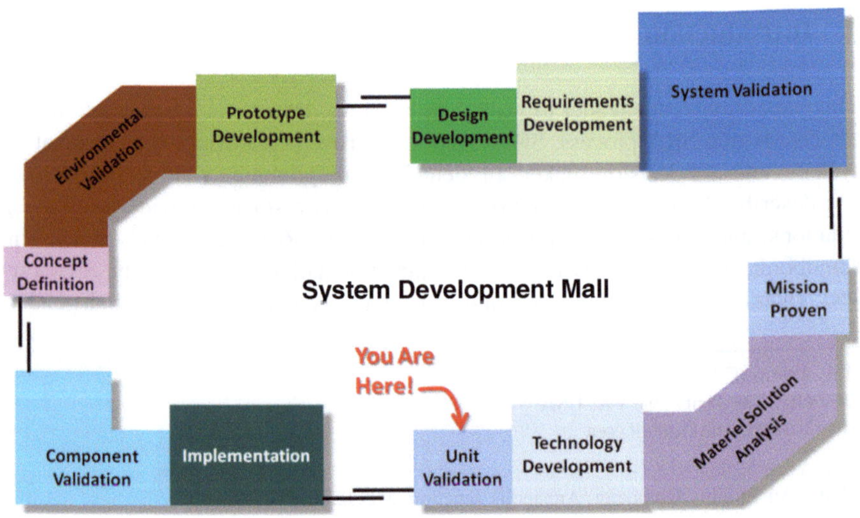

Fig. 1 The SRA indicates "you are here!" in the system development process

system-level readiness work has been adapted and enhanced from the research conducted at Stevens Institute of Technology [3]. The SRA process has been piloted on major development programs and can be applied to programs of any size enabling more effective system development management and integration that can ultimately lead to shortened delivery timelines. When applied to systems across the enterprise, SRA provides insight not only into the readiness of individual system components and functions but into capability readiness as well [4]. With a comprehensive systems view, the SRA enables developers and systems engineers to perform design trade-offs and make sound design decisions. The SRA gives the Program Manager a system perspective that allows resources to be effectively applied in the most pertinent areas.

2 Metrics

When considering product development, the integration of systems, and the portfolio management of systems, four critical questions must be answered:

1. What are we trying to accomplish? (Euphoria)
2. What can we do now? (Herding the Cats)
3. What is our plan to get there? (The Road to Euphoria)
4. How are we doing? (Product Development/Systems Acquisition/Metrics)

What we are trying to accomplish (ideal goal) can be referred to as "euphoria" [4]. The existing systems, products, and capabilities, i.e. what we can do today, are analogous to cats. The chore of the systems engineer and the program manager is to "herd the cats" such that the system development process is successfully completed deploying the required capability or function. Our plan to get there is described as the "road to euphoria". This road may take the form of an Integrated Master Schedule (IMS) or a capabilities roadmap. Finally, the program needs to know how they are doing on this road. The fourth and final question is concerned with measurement: How are we doing; what metrics can we use?

The SRA process helps us to answer both Questions No. 1 and No. 4. The core steps of the SRA process described in Sect. 3, i.e. understanding, bounding, decomposing, and mapping the system, provide the framework and information that answer Question No. 1. There are different approaches to answering the fourth question. The strength of using the SRA methodology in this instance lies in the capture of metrics that measure the vital aspects of a development effort and how well the development process is proceeding. These metrics provide input to the optimization and decision making process. Making decisions without proper systems understanding, although often done, is an inadequate approach to answering Question No. 4. Many projects do not plan properly for integration, thus incurring additional and unforeseen costs. Hugh investment decisions are made without a proper system and integration understanding. The SRA methodology provides this understanding.

An example where complete system understanding is not taken into account is current use of Technology Readiness Assessments (TRAs) by the Department of Defense (DoD). TRAs evaluate the readiness of a system as it progresses through development and the acquisition life cycle. TRAs do not assess the entire system but identify and evaluate the TRLs of Critical Technology Elements (CTEs) that are identified when performing a TRA. CTEs only capture a subjective set of "critical elements". CTEs do not cover the entire set of system and subsystem elements. TRAs do not require a comprehensive understanding of component and subsystem integration nor of other external dependencies and assessments are only performed at major milestone decisions. In many cases, the TRLs reported by the TRA are misused and misrepresented as a system level metric. While the TRA is a DoD directive and will continue to be performed across the DoD [1], the SRA process provides a "whole system" perspective that enables traceability across the entire system. The SRA is a significant enhancement that provides additional benefits not supplied by the current TRA process. The SRA:

- Measures the readiness of all system components (all elements equally critical).
- Focuses on the readiness of integration between components internal to the system and requires readiness understanding of external dependencies.
- Performs multiple analyses during the system life cycle and not just at major milestone decisions.

This section describes the five metrics used throughout the SRA process. Two of these metrics, the Technology Readiness Level (TRL) and Integration Readiness Level (IRL), are assigned. The three remaining metrics, Component SRL, Composite SRL, and SRL, are computed. Sample calculations for determining specific metrics are provided in Sect. 4.

2.1 *Technology Readiness Level (TRL)*

The TRL is a systematic metric/measurement to assess the maturity of a particular technology and allow consistent comparison between different types of technologies. TRL values range from 1 to 9 (see Fig. 2). The TRL was initially pioneered by Mankins [2] at the National Aeronautics and Space Administration (NASA) Goddard Space Flight Center in the 1980s as a method to assess the readiness and risk of space technology. Over time, NASA continued use readiness levels as part of the overall risk assessment process and as a means for comparing maturity of various technologies. NASA incorporated the TRL methodology into the NASA Management Instruction 7100 as a systematic approach to the technology planning process. The DoD, along with several other organizations including the International Standards Organization (ISO), later adopted this metric and tailored its definitions to meet their needs. A comparison of the NASA TRL definitions to the DoD and ISO [6] TRL definitions is shown in Table 1 for TRL 6.

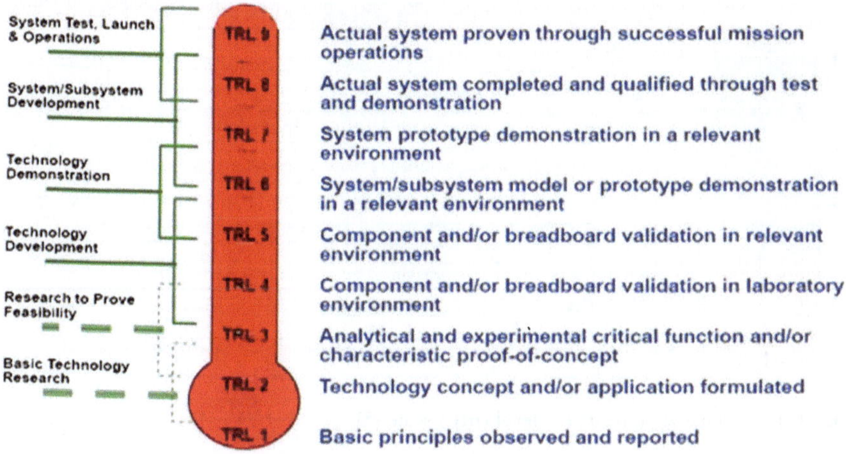

Fig. 2 Original NASA technology readiness levels (TRLs)

Table 1 A comparison of the NASA, DoD TRA guidance [1], and ISO 16290 technology readiness level (TRL) definitions for TRL 6 [6]

	NASA/defense acquisition guidebook	ISO 16290:2013(E)
TRL 6	System/subsystem model or prototyping demonstration in a relevant end-to-end environment (ground or space) System/subsystem model or prototype demonstration in a relevant environment	Model demonstrating the critical functions of the element in a relevant environment
	Prototyping implementations on full-scale realistic problems. Partially integrated with existing systems. Limited documentation available. Engineering feasibility fully demonstrated in actual system application Representative model or prototype system, which is well beyond that of TRL 5, is tested in a relevant environment. Represents a major step up in a technology's demonstrated readiness. Examples include testing a prototype in a high-fidelity laboratory environment or in simulated operational environment	Critical functions of the element are verified, performance is demonstrated in the relevant environment and representative model(s) in form, fit and function ✓ Definition of performance requirements and of the relevant environment ✓ Identification and analysis of the element critical functions ✓ Design of the element, supported by appropriate models for the critical functions verification ✓ Critical function test plan ✓ Model definition for the critical function verifications ✓ Model test reports

	IRL	Definition
Pragmatic	9	Integration is **Mission Proven** through successful mission operations.
	8	Actual integration completed and **Mission Qualified** through test and demonstration, in the system environment.
Syntactic	7	The integration of technologies has been **Verified and Validated** with sufficient detail to be actionable.
	6	The integrating technologies can **Accept, Translate, and Structure Information** for its intended application.
	5	There is sufficient **Control** between technologies necessary to establish, manage, and terminate the integration.
	4	There is sufficient detail in the **Quality and Assurance** of the integration between technologies.
Semantic	3	There is **Compatibility** (i.e. common language) between technologies to orderly and efficiently integrate and interact.
	2	There is some level of specificity to characterize the **Interaction** (i.e. ability to influence) between technologies through their interface.
	1	An **Interface** between technologies has been identified with sufficient detail to allow characterization of the relationship.

Fig. 3 Early integration readiness level (IRL) scale [5]

2.2 *Integration Readiness Level (IRL)*

The IRL is a metric to measure integration maturity between two or more components. IRLs, in conjunction with TRLs, form the basis for the SRL. IRL scale ranges from 0 to 9. The original IRL concept was developed at Stevens Institute of Technology [5] and is shown in Fig. 3. Our Vade Mecum provides an enhanced set of IRL decision criteria (Table 2) that is "evidence-based" and includes the detailed evidence description used to assess interface readiness during the SRA. The original IRL scale definitions have been modified to be consistent with the foundation of the TRL scale and to more closely reflect the indigenous development model.

IRLs characterize the systematic analysis of the interactions between various components and provide a consistent comparison of the maturity between integration points. IRLs assist the systems engineer in identifying development areas that require additional engineering. IRLs also provide a means to reduce the risk involved in maturing and integrating components into a system. Thus, IRLs supply a common comparison measure for both new system development and technology insertion.

2.3 *System Readiness Metrics*

System readiness incorporates a TRL and a metric of integration maturity, the IRL. While the TRL provides the metric for describing component knowledge, the IRL is a metric that provides a description of how well the components are integrated. System readiness provides a snapshot in time of the readiness of the entire system. There are three system readiness metrics that are computed in the SRA process.

A *Component SRL* is the System Readiness Level of an individual component of the system and its integration links. Component SRLs are used to identify which

Table 2 Enhanced decision criteria for assessing IRL

IRL	Definition	Depiction	Evidence description
0	No integration		No integration between specified components has been planned or intended
1	A high-level concept for integration has been identified		Principal integration technologies have been identified Top-level functional architecture and interface points have been defined High-level concept of operations and principal use cased has been started
2	There is some level of specificity of requirements to characterize the interaction between components	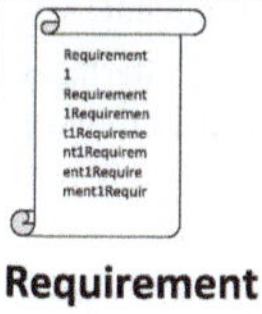	Inputs/outputs for principal integration technologies/mediums are known, characterized and documented Principal interface requirements and/or specifications for integration technologies have been defined/drafted
3	The detailed integration design has been defined to include all interface details	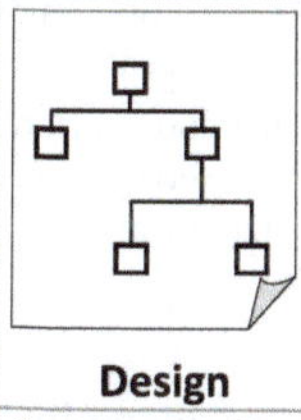	Detailed interface design has been documented System interface diagrams have been completed Inventory of external interfaces is completed and data engineering units are identified and documented
4	Validation of interrelated functions between integrating components in a laboratory environment	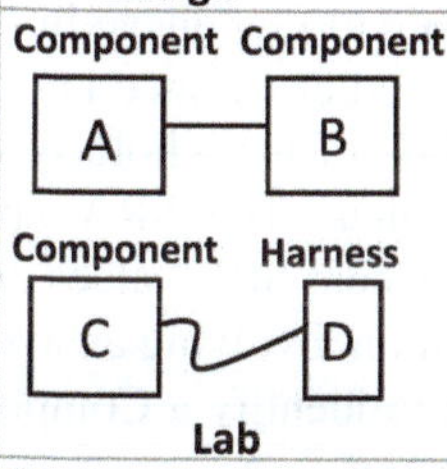	Functionality of integrating technologies (modules/functions/assemblies) has been successfully demonstrated in a laboratory/synthetic environment Data transport method(s) and specifications have been defined
5	Validation of interrelated functions between integrating components in a relevant environment	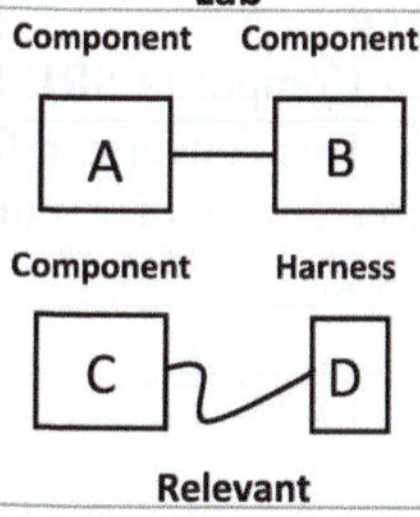	Individual modules tested to verify that the module components (functions) work together External interfaces are well defined (e.g., source, data formats, structure, content, method of support, etc.)
6	Validation of interrelated functions between integrating components in a relevant end-to-end environment	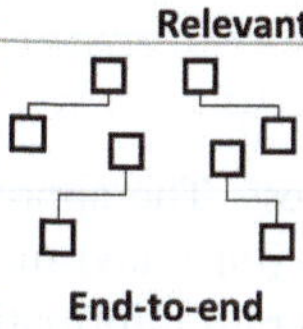	End-to-end functionality of systems integration has been validated Data transmission tests completed successfully

(continued)

Table 2 (continued)

IRL	Definition	Depiction	Evidence description
7	System prototype integration demonstration in an operational high-fidelity environment	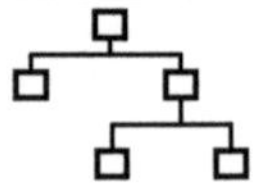 Demonstrated	Fully integrated prototype has been successfully demonstrated in actual or simulated operational environment
			Each system/software interface tested individually under stressed and anomalous conditions
			Interface, data, and functional verification complete
8	System integration completed and mission qualified through test and demonstration in an operational environment	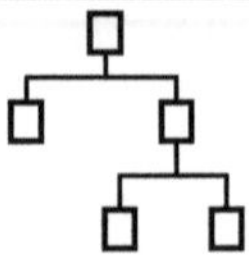 Qualified	Fully integrated system able to meet overall mission requirements in an operational environment
			System interfaces qualified and functioning correctly in an operational environment
9	System Integration is proven through successful mission-proven operations capabilities	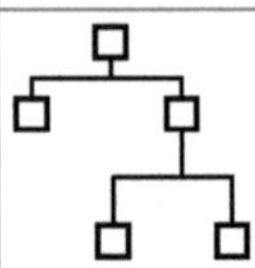 Proven	Fully integrated system has demonstrated operational effectiveness and suitability in its intended or a representative operational environment
			Integration performance has been fully characterized and is consistent with user requirement

system components may be lagging or are too far ahead in terms of their readiness and thus require Program Management and/or engineering attention.

The *Composite SRL* measures the SRL of the whole system or all of the components of the system integrated together. The SRA approach calculates the Composite SRL by averaging the Component SRL values and rendering the result in a decimal format. As with any calculation involving an average, the user needs to be aware of the potential risk of failing to identify a Component SRL that may be significantly lagging or leading the average.

The *SRL* is obtained by converting the Composite SRL to a 1–9 integer scale, with 9 being the highest level of readiness. This conversion facilitates reporting and interpreting the results, similar to the conversion of a numerical score to letter grade. This process is described in Sect. 4.2.

3 The SRA Process

This section describes in detail the SRA process. The approach for conducting an SRA is broken down into three core steps, as illustrated in Fig. 4.

The team performing the SRA gathers program information, which can include capabilities statements, requirements documents, architecture products, context

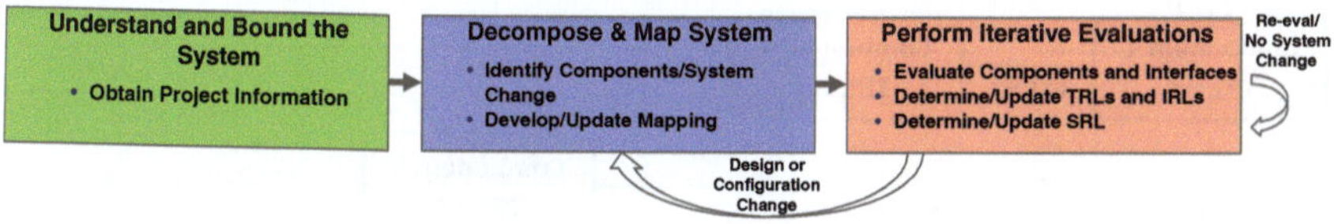

Fig. 4 System readiness assessment flow

diagrams, test plans, and any other documents that support understanding the system. Vendor product documentation and relevant published reports may provide additional information to fill any gaps for a complete understanding of the system. During this step, there is close interaction between the team, the program Lead Systems Engineer (LSE), and Subject Matter Experts.

The SRA Team uses this program information to create a mapping of the system that provides a relational understanding between the different layers of architecture. At the highest level, this mapping originates with operational requirements and activities. Functions which trace to these operational activities are then generated. System components which perform these functions are identified. The individual components are comprised of technologies. Figure 5 shows an example system mapping diagram for the components of a ten component system. This example traces from the system's operational requirements to its individual components and technologies. A component interface block diagram with ten components is then generated (Fig. 6). The SRA method is scalable to much larger systems even though the efforts to perform data gathering, assessments, and calculations increase. The system mapping and component interface diagrams serve as the foundation on which SRA analysis is performed. The system mappings identify the linkages and

Operational Activities	Service Functions (Level 1)	Service Functions (Level 2)	System Components	TRL	System Technologies	TRL
A2.1.1 Activity A2.1.2 Activity A2.1.3 Activity A2.1.4 Activity A2.1.6 Activity A2.2.1Activity A2.2.2 Activity	1.Service Function	1.1 Service Function	Component 1	4	System Technology System Technology System Technology	5 5 4
			Component 2	4	System Technology System Technology System Technology System Technology	6 4 7 5
A2.2.3 Activity A2.2.4 Activity A2.2.5 Activity		1.2 Service Function	Component 3	5	System Technology System Technology System Technology System Technology	6 5 7 5
			Component 4	7	System Technology System Technology	7 7

Fig. 5 An example mapping of a subset of a system with TRLs

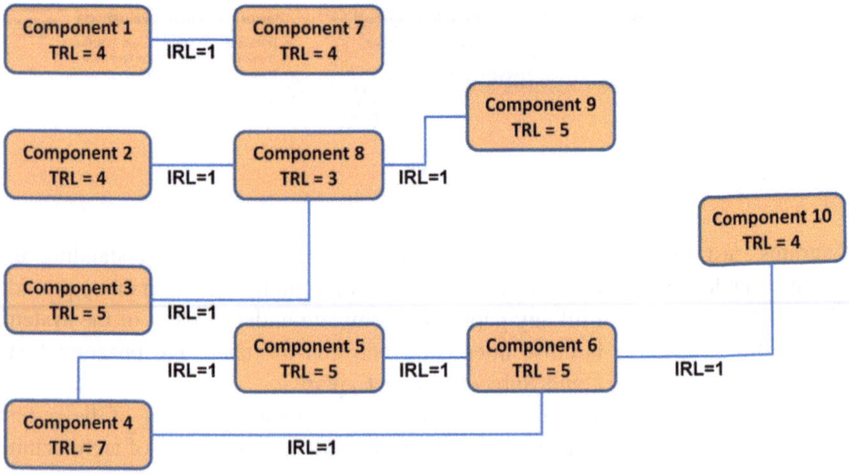

Fig. 6 TRLs and IRLs for a 10-component system

traceability between system components and allow consistent system assessments. All hardware and software components that represent the system are identified. Technologies are mapped to specific components when evaluating TRLs. Mappings are based on what is currently known and evolve and are updated as the design, architecture or other information changes. The same mapping process can be implemented when doing design or system trade-offs, providing significant benefits and insight into analysis of alternatives.

The third step in the process evaluates the system to determine its readiness. This evaluation is conducted iteratively throughout the system development cycle. The SRA process assesses readiness at three levels: Technology, Integration, and System. All component and integration links must be evaluated for technology and integration readiness. Figure 6 illustrates an example of component and integration links. TRLs and IRLs are determined using detailed decision criteria and assigned accordingly. The scales, definitions, and decision criteria of each TRL are those developed by NASA and recommended by the DoD. As shown in Fig. 5, components may be comprised of more than one technology, each with its own TRL. The TRL of a component is determined by assigning it the minimum TRL of the component's system technologies. Hence, the TRL is assessed at the technology level and the SRL is calculated at the component level. This approach for determining a component's TRL is recommended not required. Assessment may be performed at a different level as long as consistency is maintained. Figure 5 illustrates an example of a mapping and breakdown approach which carries through the sections. The SRL is then computed mathematically (see Sect. 4.1).

4 A Walk Through System Readiness Analysis

4.1 Sample Calculations

This section explains in detail and demonstrates by example the calculations and matrix mathematics used in the SRA process to determine the SRL [3]. Calculating the SRL is a function of the TRL and IRL matrices. The TRL matrix provides a snapshot in time of the state of the system with respect to the technology readiness of its components. The TRL is defined as a vector with *n* components where TRL_i is the TRL of component *i*. As discussed in Sect. 3, TRLs are mapped to specific components for evaluation purposes. The IRL matrix represents the integration of different components from a system perspective. The integration between components *i* and *j* is represented by IRL_{ij} in the IRL matrix. The theoretical integration of a component *i* to itself is denoted by IRL_{ii} and is assumed to be a maximum, i.e. 9, in this SRA approach. Zeroes in the matrix indicate no planned integration. The formation of the TRL and IRL matrices is shown in Eq. (2).

Figure 6 illustrates an example of TRLs and IRLs for a system architecture with 10 components. The components and interfaces shown have been identified using the completion of the system mapping process described in Sect. 3.

In the matrices represented, the TRL levels correspond to values 1 through 9 while the IRL values range from 0 to 9. Before performing the matrix math, these values are normalized by dividing by 9, the highest value. For example, an IRL of 9 has a normalized value of 1 for element IRL_{ij} and has the characteristics described in Table 2 with respect to the *ith* and *jth* components. Similarly, an IRL of 5 has a normalized value of $\frac{5}{9}$ or 0.556.

At a minimum, each of the components in a system is connected to one other component. This integration is bi-directional, and it is assumed that the IRL is the same in each direction. Each component is integrated with other components in a specific way and is used to formulate and calculate the SRL. A zero (0) is placed in the matrix where no integration is planned.

In order to calculate a value of the SRL an SRL matrix is generated by obtaining the product of the IRL and TRL matrices, as shown in Eq. (1):

$$[SRL]_{nx1} = [IRL]_{nxn} x\ [TRL]_{nx1} \tag{1}$$

The SRL matrix consists of one element for each of the constituent components and, from an integration perspective, quantifies the readiness level of a specific component with respect to every other component in the system while also accounting for the development state of each component.

$$\begin{pmatrix} SRL_1 \\ SRL_2 \\ SRL_3 \\ \ldots \\ SRL_{10} \end{pmatrix} = \begin{pmatrix} IRL_{1,1} & IRL_{1,2} & \ldots & -IRL_{1,10} \\ IRL_{2,1} & IRL_{2,2} & \ldots & -IRL_{2,10} \\ \ldots & \ldots & IRL_{3,3} & -\ldots \\ \ldots & \ldots & \ldots & -\ldots \\ IRL_{10,1} & IRL_{10,2} & IRL_{10,3} & -IRL_{10,10} \end{pmatrix} \times \begin{matrix} TRL_1 \\ TRL_2 \\ TRL_3 \\ \ldots \\ TRL_{10} \end{matrix} \quad (2)$$

Mathematically, for a system with $n = 10$ components, the SRL is as shown in Eq. (2), where TRL_i represent the individual TRLs and the IRL_{ij} are the individual IRLs between the components. SRL_i represents the readiness level of Component i, reflecting the readiness of *all* of its connections/interfaces. (Recall that IRL_{ij} represents the IRL only between Component i and Component j.)

The corresponding SRL_i for each component i is then divided by m_i, as shown in Eq. (3), to obtain its normalized value. The m_i term is the number of integrations of component i with every other component as defined by the system architecture. This includes integration of the component with itself. Dividing by m_i also allows each component to be neutrally weighted looking at the component in isolation with its nearest neighbors, yielding well-behaved and consistent mathematical and statistical properties.

$$Component\ SRL_i = \frac{SRL_i}{m_i} \quad (3)$$

The Composite SRL for the system is the average of the Component SRL values, as shown in Eq. (4), where n is the number of components:

$$Composite\ SRL = \frac{\left(\frac{SRL_1}{m_1}\right) + \left(\frac{SRL_2}{m_2}\right) + \left(\frac{SRL_3}{m_3}\right) + \ldots + \left(\frac{SRL_{10}}{m_{10}}\right)}{n} \quad (4)$$

4.2 Results and Interpretation

Performing the Systems Readiness Assessment for the ten component system shown in Fig. 6 and working through the matrix algebra yields the resultant ten Component SRLs as shown in Table 3. The Component SRLs are important, as they provide an indicator of the readiness of the individual components and their associated integrations. Comparing individual Component SRL values relative to each other identifies those components that are lagging or may be too far ahead in their "readiness." For example, from Table 3, Component 8s SRL is lagging the

Table 3 Component SRLs for 10 component system

1	2	3	4	5	6	7	**8**	9	10
0.247	0.241	0.296	0.300	0.235	0.163	0.185	**0.127**	0.176	0.253

other system components as it has a much lower Component SRL value of 0.127. This is brought to the attention of the decision makers for a risk assessment and potential further analysis.

The Composite SRL is the average of the Component SRLs (Eq. 5). When averaging, the user needs to take into account the potential risk of masking a Component SRL that is significantly lagging or leading the average, reiterating the importance of assessing and monitoring the individual Component SRLs.

$$\begin{aligned}\text{Composite SRL} &= [0.247 + 0.241 + 0.296 + 0.300 + 0.235 + 0.163 + 0.185 \\ &\quad + 0.127 + 0.176 + 0.253]/10 \\ &= 0.222\end{aligned} \tag{5}$$

Composite SRLs are defined on a scale from 0 to 1 with the value carried out to three decimal places. For the calculations in the example above, the Composite SRL is reported as 0.222 with the 10 Component SRLs shown. It could potentially be difficult to understand the difference between system readiness values that are very similar (e.g., 0.247 vs. 0.241 vs. 0.296). Composite SRL values are translated to whole numbers consistent with TRL and IRL scaling for ease of interpretation. To translate the 0 to 1 scale to a 1 to 9 scale, the SRL Translation Model shown in Table 4 is used to map the decimal values to whole numbers. Because the SRA is dependent on the system architecture, a SRL Translation Model is generated for each architecture configuration when performing the SRA as shown in Table 4.

To generate the SRL Translation Model for this example architecture, a Composite SRL_i is calculated for nine system architecture configurations (each with 10 components and 10 integration links) where the TRLs for all of the components

Table 4 SRL translation model

TRL	IRL	Composite SRL_i	Midpointbetween levels	Composite SRL_i	SRL
9	9	1.000		0.914–1.000	9
			0.914		
8	8	0.828		0.750–0.913	8
			0.750		
7	7	0.672		0.601–0.749	7
			0.601		
6	6	0.530		0.467–0.600	6
			0.467		
5	5	0.404		0.349–0.466	5
			0.349		
4	4	0.293		0.245–0.348	4
			0.245		
3	3	0.197		0.157–0.244	3
			0.157		
2	2	0.116		0.084–0.156	2
			0.084		
1	1	0.051		0.000–0.083	1

Table 5 SRL descriptions[a]

9	System has achieved initial operational capability and can satisfy mission objectives
8	System interoperability should have been demonstrated in an operational environment
7	System threshold capability should have been demonstrated at operational performance level using operational interfaces
6	System component integrability should have been validated
5	System high-risk component technology development should have been complete; low-risk system components identified
4	System performance specifications and constraints should have been defined and the baseline has been allocated
3	System high-risk immature technologies should have been identified and prototyped
2	System materiel solution should have been identified
1	System alternative materiel solutions should have been considered

[a]Derived from the DoD integrated defense acquisition, technology and logistics life cycle management system chart

and the IRLs for all of the integration links are set equal to the same value, an integer from 1 to 9. For example, the Composite SRL of 0.051 is calculated by setting the TRL of each of the 10 components equal to 1 and the IRL of each of the 10 integration links equal to 1. The midpoints between each pair of adjacent Composite SRL_i are used as the boundaries for the corresponding Composite SRL_i Range values, as shown in Table 4.

The complete SRL scale and descriptions are given in Table 5. The SRA shown in the example at the beginning of this section resulted in a Composite SRL of 0.222. Using the SRL Translation Model, this translates to a System Readiness Level of 3. This indicates that immature and high-risk technologies have been identified and prototyped. Potential concerns are documented, and reported.

The SRL calculated in this example represents a snapshot in time. It is critical to measure the system readiness at multiple points along the life cycle to avoid pitfalls that can occur when readiness is only assessed only once or twice. The SRA for this example was conducted early in the life cycle. TRLs and IRLs of the components progress with system development. The additional snapshots of the SRA provide the decision maker with better risk assessment information.

SRAs can be performed on any size program and at any time during system development. The potential technology and integration risks determine the frequency at which SRAs should be performed. For larger programs, a quarterly SRA is recommended while for small programs SRAs may be performed every month. Once the system has been defined, the system mapping completed, and the initial SRA done, subsequent SRAs can be performed in a reasonably short time.

5 Guidelines for Successful Implementation of the SRA Process

In order to properly and effectively use the SRA Process, we provide the following set of guidelines.

When no integration is planned between two components the IRL value is set equal to 0. If integration between components is planned but not yet established, set the IRL value equal to 1. Avoid interim or nodal comparison of TRLs and/or IRLs that result in setting an expectation for what the aggregate readiness/maturity should be. Let the SRL approach "work for itself." Do not solely use the SRL, a single number, as the basis for decision making but rather take into consideration all the lower level metrics, e.g., Component SRLs. Use these Component SRLs to identify components or areas of systems development that are lagging or too far ahead in their readiness progression.

The SRL should be used as an indicator of current system readiness rather than for predictive analysis. The intent of the SRL approach is not to estimate "how long" nor does it measure the level of effort it takes to increase system readiness. Only compare SRLs of the same system throughout its life cycle. Compare "your system" as it matures, not two different systems. The SRL is a comprehensive snapshot in time of the readiness of the current architecture of the system and can be used as a valid indicator of readiness for the system at that time. The calculation of the SRL is dependent on the structure of the system. Adding components and/or interfaces changes the structure of the system and a careful examination should be performed to ensure that inconsistencies are not introduced.

6 Conclusion

SRAs are an innovative methodology that provides system level metrics to help reduce integration issues, a leading cause of system development failures. The SRA methodology provides decision-makers with a snapshot of a system's holistic state of maturity and quantifies the level of component-to-component integration during system development, helping to improve system performance management. Implementation of the SRA methodology aids decision makers in identifying both programmatic and technical risk areas. A number of program pilots currently validate the SRA methodology with further validation to be achieved through application across multiple enterprises. Future research includes mathematically sound weighting techniques and leveraging the principles of the SRA framework to model system availability and for other Risk Management techniques.

References

1. Assistant Secretary of Defense for Research and Engineering (ASD(R&E)), Department of Defense Technology Readiness Assessment (TRA) Guidance, April 2011
2. Mankins, J.C.: Technology Readiness Levels, NASA (1995)
3. Sauser, B., Ramirez-Marques, J., Magnaye, R., Tan, W.: A systems approach to expanding the technology readiness level within defense acquisition. Int. J. Defense Acquisition Manage. **1**, 39–58 (2008)
4. Austin, M.F., et.al.: A systems approach to the transition of emergent technologies into operational systems—herding the cats, the road to euphoria and planning for success. In: INCOSE International Symposium, Utrecht, Netherlands, 2008
5. Gove, R.: Development of an integration ontology for systems operational effectiveness. M.S. Thesis, Stevens Institute of Technology, Hoboken, NJ (2007)
6. ISO/FDIS 16290:2013-11 (E). Space systems—definition of the technology readiness levels (TRLs) and their criteria of assessment

Designing and Integrating Complex Systems: Be Agile Through Liveness Verification and Abstraction

Thomas Lambolais, Anne-Lise Courbis, Hong-Viet Luong and Thanh-Liem Phan

Abstract Model Driven Architecture (MDA) is recognised as a strong way to develop high-quality systems, and specifically reactive systems. Within MDA, models are in the center of a stepwise development based on extensions, refinements and transformation. Systems Engineering addresses the problem of complex system development in a holistic way, however, there is a lack of tools to verify models from a behavioural point of view at the earlier stage of the development, taking into account that the specifications are evolving during the system development. We propose IDF, a framework for Incremental Development of Compliant Models, which is constituted with a set of relations based on the verification of liveness properties. It is computed on abstract models automatically set up from behavioural specifications of the system or its component. These relations detect non-conformance of models during their evolution (extension or refinement) such as the non-interoperability of sub-components belonging to an architecture.

1 Introduction

Model Driven Architecture (MDA) [1] is recognised as a strong way to develop high-quality systems, and specifically reactive systems which are event-driven systems that must continuously react to external stimuli. Such systems include for

T. Lambolais · A.-L. Courbis (✉)
LGI2P école des mines d'Alès, Site de Nîmes, Parc Scientifique Georges Besse,
30 035 Nîmes cedex 1, France
e-mail: anne-lise.courbis@mines.ales.fr

H.-V. Luong
M2 M-NDT, 1 Rue de Terre Neuve, Miniparc du Verger, bâtiment H,
91 940 Les Ulis, France

T.-L. Phan
LSEI, CEA INES, 50 Avenue du lac Léman, BP 258,
73 375 Le Bourget du Lac Cedex, France

G. Auvray et al. (eds.), *Complex Systems Design & Management*,
DOI 10.1007/978-3-319-26109-6_5

instance embedded controllers for automotives, avionics, train, telephony, but also communication network.

Within MDA, models are in the center of a stepwise development based on model extensions, refinements and transformations, from an abstract incomplete specification to a concrete complete model. By this way, models serve both as a description of the problem domain, i.e. a requirement, and a specification for the implementation, bridging the gap between problem and solution. Many methods and tools have been proposed to support model development based on standard modelling languages such as UML or SysML. Methodologies are also necessary in order to deal with complex systems. Systems Engineering [2] addresses this challenge in a holistic way considering both business and technical aspects of a system design, integrating all stakeholders at the early stage of the development, starting from the user requirements and the definition of the environment of the system to be designed in order to produce high-quality systems. Many methodologies and many standards have been proposed to follow these recommendations as it is shown in the survey proposed in [3]. Our area of interest focuses on the definition and the analysis of the behavioural view of the system, expressed by a functional or organic architecture whose components are defined by a behavioural view or an architectural one. The target activities are therefore the functional analysis, the functional verification and the synthesis in the IEEE 1220 Process model [4]. Our experience in system modelling highlighted that architecture definition, behavioural abstraction and refinement are the core activities of system design. Designing a system consists not only in modelling its architecture, but also in evaluating its behavioural models and that of its components at the beginning of the modelling process, although the model is incomplete and non-deterministic. These features have to be considered as a support for designers and architects. It means that such verifications have not to be postponed at the end of the modelling process. They have to be integrated in the incremental development of the system and its components.

For this propose, we have defined IDF, an Incremental Development Framework. It is defined by a set of relations computed on an abstract formalism (LTS for Labelled Transition System), allowing models to be evaluated during their development. The environment of the system to be designed can be at its turn modelled taking into account its uncertain or non-deterministic behaviour. By this way, incompatibility or non-interoperability can be detected at early stages of the design process. The framework is supported by a tool, named IDCM (Incremental Development of Compliant Models). Experiments have been conducted on UML models. Our work is inspired by techniques of model checking [5]. Such verifications aims at:

- supporting the stepwise realisation of systems by applying refinement and extension operations
- analysing the interaction of the system with its environment, with respect to non-deterministic scenarios
- insuring the interoperability of the system components
- insuring the evolution of the system by substituting a component by a new one

This paper gives an overview of the concepts of IDF and tools we have developed to support IDF. The following section presents modelling concepts of architectures and behavioural components through an incremental development process in order to point out topics being addressed. Section 2 introduces definition of liveness and abstraction models allowing UML/SysML models to be analysed. Section 3 gives an overview of relations we have implemented to support IDF. Section 4 shows main functionalities of the tool IDCM for supporting IDF concepts. A presentation of our future work will close this article.

2 The Architectural Paradigms

In this section, we present main useful concepts to understand our proposal for incremental development of architectural models. We focus on the verification of behavioural specifications of a system all along its design life cycle. Figure 1 gives an overview of the useful operations for the development of a system based on a MDA approach. We suppose that the first step starts by defining a behavioural specification of the system (BEHAV1 in Fig. 1) at a high abstraction level. Such a specification may evolve and be extended (BEHAV2 in Fig. 1) until an agreement is reached between the various stakeholders of the system development (client, end-users, designers). This agreement may however evolve during the system

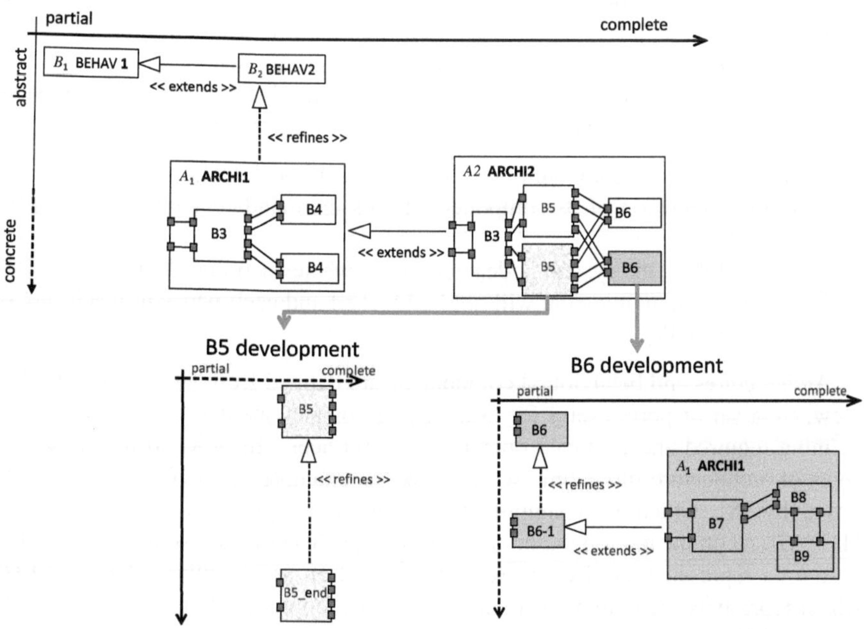

Fig. 1 Overview of an incremental development through refinement and extension operations

design process and at every step, it will be necessary to be able to take into account new specifications.

When the system is complex, its design is structured into components that may represent functional components or physical components depending on the stage of the design process. Components defined according to a structural view are called architectures. For example, in Fig. 1, the first architecture is named ARCHI1; it is extended into ARCHI2 whose components have to be refined. Architectures can be seen as a hierarchical tree whose leaves are behavioural components. Architectures may represent logical architectures or physical ones.

Extensions means that new behaviours are introduced into the design, for whatever reasons: the system is too complex to be defined in one shot, the client changes is mind, there is an already developed COTS whose specification is closed of the required one that could be integrated with lower cost, a product line has already be tested and its enhancement is expected by introducing new requirements, and so on.

Refinements aim at adding details and reducing non-determinism in order to get a concrete model closer to the final implantation of the system.

Developments of components may be processed by separate teams, by means of a collaborative platform, that increase the complexity of the process. One main concern of component designers is to develop components that meet their specification. Components are supposed to be defined for a given context, except that this context is evolving since it is itself under development. One goal of the architect is to verify the behavioural consistency of the models being developed. This task is critical since sub-systems have their own development life cycle. Nevertheless, the architect cannot wait until the final implantation model to check the consistency analysis of the system. He/she has to maintain the functional consistency of the system model under development whatever the abstractions of sub-system models. We characterize consistency by the following properties:

- conformance: the behavioural specification of the architecture that is deduced from the interaction of its components fulfils the mandatory parts of the specification [6].
- interoperability: the system is deadlock free; whatever point of interaction may be reached, communication will not be blocked and each part will reach one of its final states [7].

Architectures and behavioural components are defined from an external point of view, by a set of ports useful for establishing connections and a set of interfaces defining required and provided operations (or services). In order to illustrate concepts of architecture modelling, we will take as example the V76 case study proposed by [8], which is a simplified version of the protocol described in the ITU V.76 recommendation, based on LAPM (Link Access Procedure for Modems). Figure 2a represents an abstract external view of an architecture named V76-DL which represents the communication between two components that implement the protocol V76 and Fig. 2b is a more detailed external view.

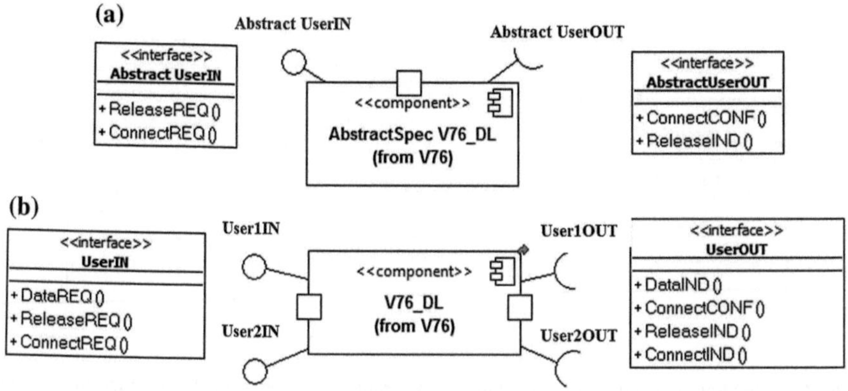

Fig. 2 External view of two points of view of architecture V76-DL

The internal view of an architecture is defined by its components and their interconnections. For example, Fig. 3 illustrates the internal view of architecture V76-DL: it is constituted with two components of type V76 whose external view is given in Fig. 4. The architecture allows two users to communicate through the ports U1 and U2.

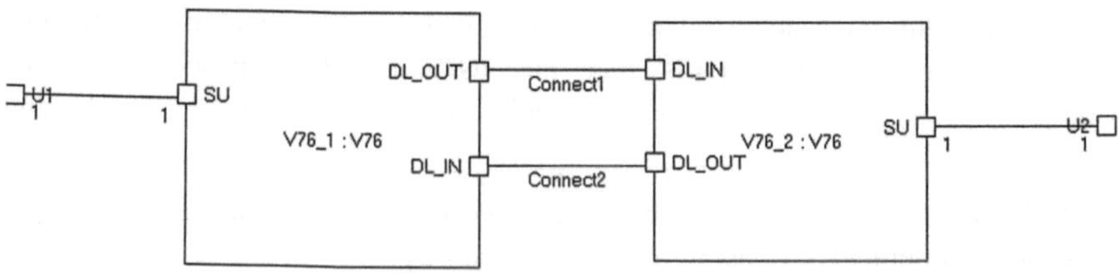

Fig. 3 Internal view of architecture V76-DL

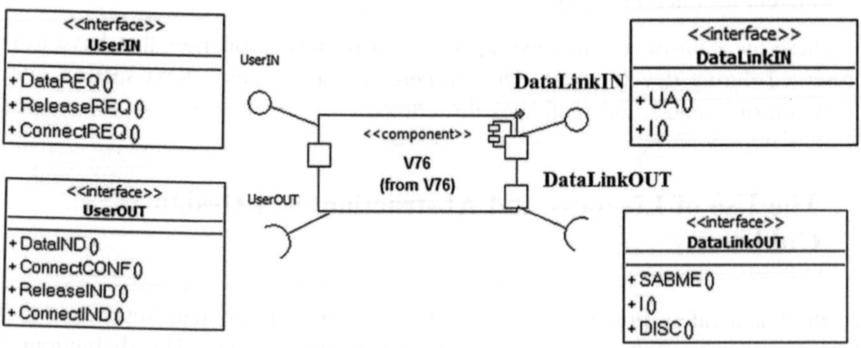

Fig. 4 External view of component V76

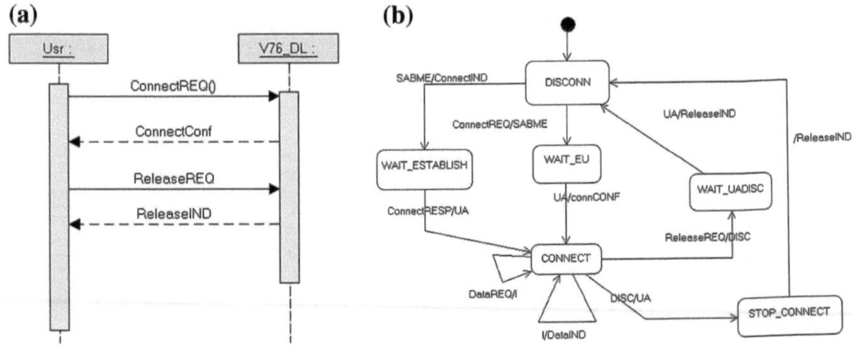

Fig. 5 Behavioural specifications: **a** sequence diagram associated with the abstract architecture V76-DL, **b** state machine of component V76

The internal view of a behavioural component is defined by a behavioural specification defined according to its ports, the operations of its external view and private internal operations. Many formalisms may be used for behavioural specification depending on the system features and the progress of the development: sequence diagrams, state machines, functional flow block diagrams. For example, Fig. 5a shows a simplified specification of the architecture V76-DL from the transmitting user point of view and Fig. 5b shows the state machine of component V76 belonging to architecture V76-DL.

Analysing the consistency of an architecture during its development requires specific mechanisms and tools that are usually not proposed by CASE (Computer-Aided Software Engineering) tools. These mechanisms are divided into two groups:

- model verifications: adequate relations have to be defined to capture conformance, refinement, extension and interoperability
- model abstraction: adequate models have to be set up from the model under construction in order to capture behavioural specification from an external point of view and an appropriate abstraction in order to compare models defined at different abstraction levels.

These mechanisms are defined according to liveness properties that have to be preserved during development. This property is the liveness. Next section gives definition of liveness and motivates this choice.

3 The Use of Liveness and Abstraction as a Design Guideline

Liveness and safety properties allow systems to be analysed with respect to their behavioural specification as observed by their environment. This behaviour is observed by traces which are partial sequences of interactions (events or actions) starting from the initial state of the system. There are several ways to define safety and liveness, some of them being contradictory about the classification of deadlock property. We have selected definitions proposed by [9]: a safety property asserts that the system always stays within some allowed set of finite behaviours, in which nothing "bad" happens. The violation of such properties occurs after a finite execution of the system. A liveness property asserts that the system eventually reaches a good set of states, that means it will eventually react as it should after some given traces. A liveness property represents what the system must do, while a safety represents what the system has not to do. When reasoning on models, liveness properties can only be established under some fairness assumption, stating that the system is not allowed to continuously favour certain choices at the expense of others [10]. The fairness assumption implies that the system will eventually accept an event occurring infinitely often. Lastly, we consider that deadlock freedom is a liveness property, as proposed in [11] since a deadlock means that the system refuses any input event.

Many formal methods addressing complex system development advocate refinement techniques [12, 13] such as B method [14] or Object-Z [15]. They focus on the preservation of safety properties all along the process of development. Such methods are adequate when the specification of the component or the complete system is definitive and not being defined or evolved. Another way to support designers during model development is to preserve the liveness properties as mentioned in [16]: liveness properties act as a design guideline for developing systems.

Liveness is crucial for reactive systems and is complementary to safety to support designers during an incremental development: observing liveness allows specification to be enriched, starting from a "draft" model that is completed by a stepwise approach in a non-regressive way.

It is therefore necessary to provide designers with tools to compare models according to their liveness properties, taking into account that they sub-components can be defined at different abstraction levels. For example, how ensuring that architecture V76-DL fulfils the behavioural specification expressed by the sequence diagram? Are components of architecture V76-DL interoperable?

To answer these questions, we have defined two mechanisms: model abstraction and model analysis based on a liveness analysis.

3.1 Model Abstraction

With model abstraction, a simplified behaviour is extracted from models to be analysed. This extraction takes into account several criteria: the abstraction levels of models to be compared, the type of relation to be analysed (extension, refinement or interoperability), and of course, the goal of the analysis that is based on the analysis of the interaction of system (or one of its sub-system) and its environment. Abstract models are formalised by LTS (Labelled Transition System) [17]. Reasoning on such a formalism has many advantages: the system analysis is independent from the modelling formalism chosen by the designer; models can thus be compared even if their application domain is different, that is usual in System Engineering; existing relations already defined on LTS can be used for our purpose.

We do not formally introduce LTS and the process to abstract state machines into LTS. You can refer to [18] and [19] to get details about the transformation. Figure 6a illustrates the LTS generated from the state machine of component V76, and Fig. 6b the LTS associated with the sequence diagram of the architecture V76-DL. The transformation does not handle data; it only focuses on provided and required events (or services) offered by the component under analysis. When the component is an architecture, we have defined a transformation [20] which computes all combinations of internal events between components and reduces the LTS to observable events by hiding internal synchronisations and internal operations. Hidden actions are noted i in the LTS. For example, the LTS associated with the architecture of Fig. 3 handles operations defined on its interfaces given in Fig. 2b. Operations defined on interfaces of internal components, that is interfaces DataLinkIN and DataLinkOUT, are hidden. The LTS is built by synchronising the two LTS of Fig. 6 on their internal connector. It contains 84 transitions and 54 states.

When models to be compared do not belong to the same abstraction level, their interfaces may be different. For example, there are more operations in interfaces of

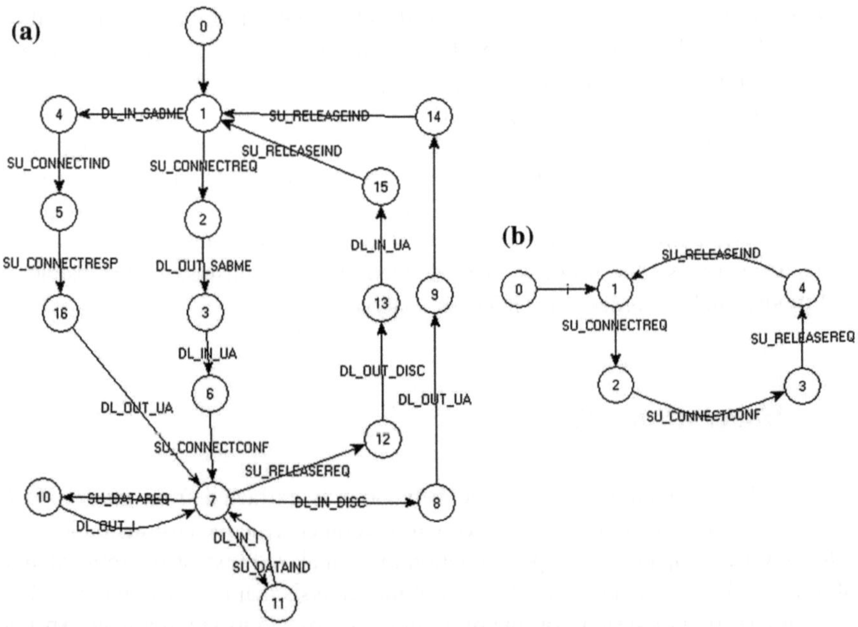

Fig. 6 **a** LTS associated with the state machine of component V76, **b** LTS associated with the sequence diagram of the simplified specification of architecture V76-DL

component V76-DL than those of the specification of V76 protocol given by the sequence diagram. Comparison needs to align the abstraction levels. For this purpose, we use a hiding mechanism and a renaming mechanism, when operations are refined. For example, to compare V76-DL and the sequence diagram, internal operations of the architecture (`ua`, `i`, `sabme`, and `disc`) are hidden such as the operations belonging to the port `u2`, which correspond with the user receiving the data. By this mechanism, the LTS associated with V76-DL architecture will be comparable to the abstract specification.

The main feature of this abstract model is that it captures what the system must do and what the system may do. That is crucial for liveness properties as we point out below.

3.2 Liveness Analysis

There exists a specific relation, which lonely goal is to preserve liveness. This relation is conformance relation `conf` [21, 22]. Conformance testing methodologies proposed by ISO and ETSI [6] are designed to compare an implementation model with a standard specification. Standard specifications or recommendations serve to define both the mandatory and optional parts. The main idea behind conformance is to verify agreement between an implementation and its specification on required parts; informally speaking, an implementation conforms to a standard if it has properly implemented all *mandatory parts* of the standard [23].

For instance, in Fig. 7, we can deduce the following properties:

- *spec*1, *spec*2 and *spec*4 may accept `releaseREQ` or `connectREQ` after a sequence of `connectREQ`. As they may also refuse them, operations `releaseREQ` or `connectREQ` are optional.
- *spec*3 must accept `releaseREQ` after `connectREQ`. `releaseREQ` is thus mandatory after the trace `connectREQ`.

We can verify: *spec*1 `conf` *spec*2, *spec*2 `conf` *spec*1, *spec*1 `conf` *spec*4. However, *spec*1 ¢onf *spec*3: from an observational standpoint, nothing distinguishes *spec*1 from *spec*3 but `conf` relation detects non-determinism of *spec*3. In this example, *spec*1 may refuse `releaseREQ` after a non-empty unbounded occurrences of `connectREQ`, whereas *spec*3, which is deterministic, cannot. *spec*1 and *spec*3 are trace equivalent, yet not in conformance. Lastly, even if *spec*1 `conf` *spec*4 and *spec*4 `conf` *spec*1, we can verify that *spec*4 cannot substitute *spec*1.

Even though the conformance relation has been defined by [22], we are still not aware of any published method to compute it. We have thus proposed an implantation of this relation and pointed out how extension and refinement relations can be defined from the conformance relation [19, 24]. In the same way, we have implemented the procedure allowing to check if a component can substitute another one, whatever its environment may be [20].

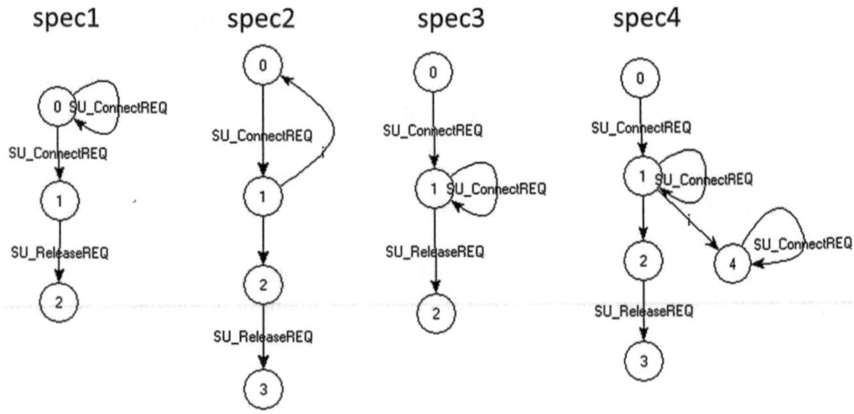

Fig. 7 Example of `conf` relation

Next section gives an overview of the tool IDCM we have defined and implemented to provide designers with a tool box to analyse models.

4 IDCM: Incremental Development of Compliant Models

IDCM is a tool box allowing models to be compared with respect to refinement, extension and substitution relations. It is based on concepts of IDF focusing on the analysis of liveness properties and abstraction of behavioural/functional models. It is developed in Java. Its first release is integrated into TopCased environment [25] and focus on UML state machines and composite component analysis. When a model is loaded for verification, the set of its components is proposed to be abstracted into LTS (see Fig. 8).

Behavioural component transformation is performed by an ad hoc algorithm we have developed by parsing state machine xmi models. Composite components transformation is done with two stages: the first one produces an intermediate file in EXP.OPEN format [26] that is obtained by parsing composite component xmi models; the second stage, consisting in transforming the intermediate file into LTS, is performed by the CADP toolbox [27]. LTS associated with state machines and composite components are generated into CADP textual and binary formats [27].

IDCM proposes a set of relations for model comparison. They are classified in several families: relations for incremental development (extension or refinement), relation for liveness verification to check the conformance between an implantation and its specification, relations for assembling sub-components (compatibility) and lastly, relations to check if a component can substitute another one. When a relation between two models does not hold, a verdict is given as a sequence of observable

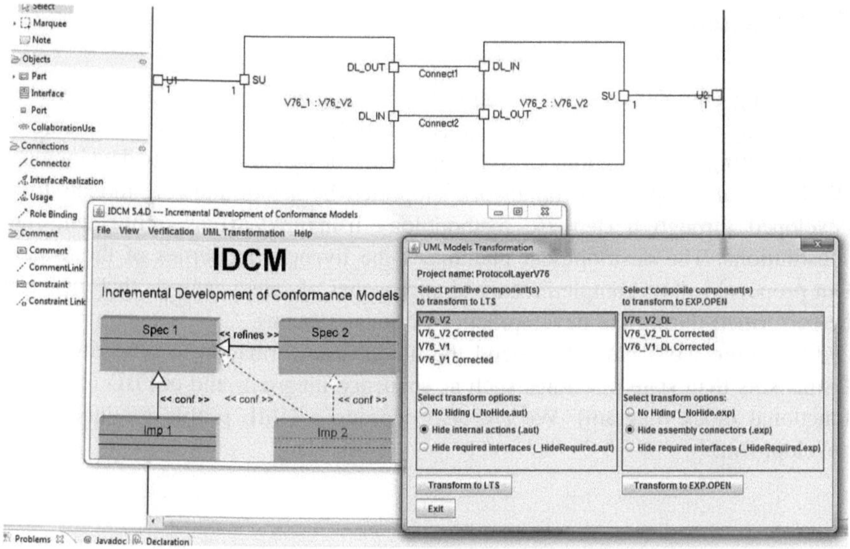

Fig. 8 Interface to transform behavioural and architectural components into LTS

events leading to a failure. Designers are in charge to analyse the trace, to execute it on the state machine, or in the architecture in order to find the mistake and correct it. For example, we have found a mistake (Fig. 9) in the state machine of component V76 by comparing the architecture with its abstract specification. There exists a deadlock after the action `connectREQ` when the two users send together a `connectREQ`. We have corrected this mistake by adding a state and transitions between `wait-eu` and `wait-establish` states in the state machine of Fig. 5b.

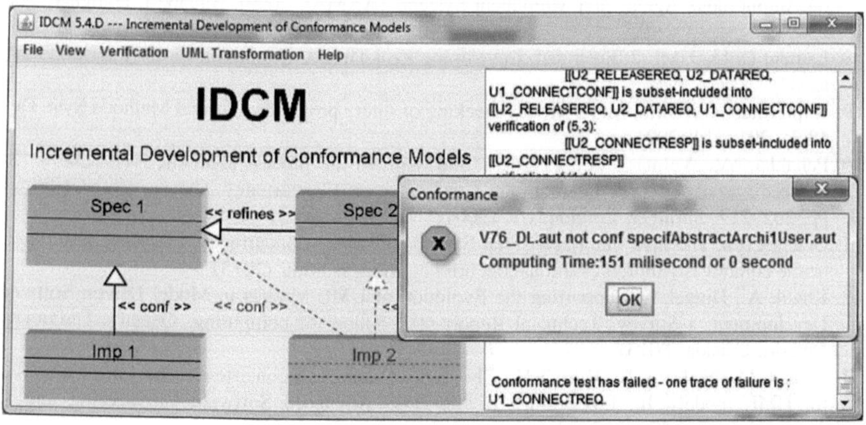

Fig. 9 Verdict of the conformance between the architecture V76-DL and its abstract specification

5 Conclusion

Developing complex systems requires methodologies such as MDA and System Engineering. Nevertheless, there is an actual difficulty for designers and architects for evaluating the behaviour of a system being designed during its development. We have thus proposed a framework supported by a tool allowing models to be developed through a stepwise methodology using extensions, refinements and substitutions. The development guarantees the liveness properties of the system. Our proposal is thus complementary to approaches of safety analysis that must also be performed during the development of critical systems.

Our future work plans to extend the model transformation to other functional formalisms than state machines such as sequence diagrams and eFFBD (enhanced functional block diagram). We are also defining a UML profile for incremental development.

References

1. OMG MDA. Model Driven Architecture Foundatation Model. OMG ormsc/10-09-06 (2006)
2. Systems engineering handbook. INCOSE (2006)
3. Estefan, J.A.: Survey of model-based systems engineering (mbse) methodologies. Technical Report INCOSE-TD-2007-003-01, INCOSE MBSE Focus Group (2008)
4. IEEE 1220-2005. Standard for application and management of the systems engineering process. In: IEEE Computer Society (2005)
5. Clarke, E.M.: The birth of model checking. In: 25 Years of Model Checking. Lecture Notes in Computer Science, vol. 5000, pp. 1–26 (2008)
6. ISO/IEC9646. Information technology—open systems interconnection—conformance testing methodology and framework—part 1: general concepts (1991)
7. Baldoni, M., Baroglio, C., Chopra, A.K., Desai, N., Patti, V., Singh, M.P.: Choice, interoperability, and conformance in interaction protocols and service choreographies. In: Sierra, C., Decker, K.S., Sichman, J.S., Castelfranchi, C. (eds.) 8th International Conference on Autonomous Agents and Multiagent Systems (AAMAS 2009). Budapest, Hungary, May 2009
8. Laurent Doldi. UML 2 Illustrated: Developing Real Time & Communication Systems. TMSO (2003)
9. Kupferman, O., Vardi, M.Y.: Model checking of safety properties. Formal Methods Syst. Des. **19**(3), 291–314 (2001)
10. Puhakka, A., Valmari, A.: Liveness and fairness in process-algebraic verification. In: Proceedings of the 12th International Conference on Concurrency Theory, CONCUR '01, pp. 202–217. Springer, London, UK (2001)
11. Oracle Corp. The Java Tutorials—Trial Essential Classes: Concurrency. Liveness. http://docs.oracle.com/javase/tutorial/essential/concurrency/liveness.html/ (2015)
12. Khalil, A., Dingel, J.: Supporting the Evolution of UML Models in Model Driven Software Development: a Survey. Technical Report 602, School of computing, Queen's University, Ontario, Canada (2013)
13. Usman, M., Nadeem, A., Kim, T.H., Cho, E.S.: A survey of consistency checking techniques for UML models. In: Proceedings of the 2008 Advanced Software Engineering and its Applications, pp. 57–62 (2008)

14. Abrial, J.-R.: Modeling in Event-B—System and Software Engineering. Cambridge University Press, Cambridge (2010)
15. Smith, G.: The Object-Z Specification Language, Volume 1 of Advances in Formal Methods. Kluwer Academic Publishers, Boston (2000)
16. Hudon, S., Hoang, T.S.: Systems design guided by progress concerns. In: Integrated Formal Methods, pp. 16–30. Springer, Berlin, Heidelberg (2013)
17. Milner, R.: Communication and Concurrency. Prentice-Hall, Inc., New York (1989)
18. Lambolais, T., Courbis, A.-L., Luong, H.-V., Phan, T.-L.: Interoperability analysis of systems. In: 18th World Congress of the International Federation of Automatic Control (IFAC 2011), pp. 7879–7884 (2011)
19. Luong, H.-V.: Construction incrémentale de spécifications de systèmes critiques intégrant des procédures de vérification. PhD thesis, Université Paul Sabatier Toulouse III, Oct 2010
20. Phan, T.-L.: Développement incrémental de spécifications d'architectures en UML intégrant des procédures de vérification. PhD thesis, Université Montpellier II (2013)
21. Cleaveland, R., Steffen, B.: A preorder for partial process specifications. In: CONCUR '90 Theories of Concurrency: Unification and Extension, pp. 141–151. Springer, New York, NY, USA (1990)
22. Leduc, Guy: A framework based on implementation relations for implementing LOTOS specifications. Comput. Netw. ISDN Syst. **25**, 23–41 (1992)
23. Moseley, S., Randall, S., Wiles, A.: In pursuit of interoperability. In: Jakobs, K. (ed.) Advanced Topics in Information Technology Standards and Standardization Research, Chap. 17, pp. 321–323. Idea Group Publishing, Hershey (2006)
24. Luong, H.-V., Lambolais, T., Courbis, A.-L.: Implementation of the conformance relation for incremental development of behavioural models. In: Czarnecki, K. (ed.) Proceedings of 11th International Conference on Model Driven Engineering Languages and Systems (MoDELS). Lecture Notes in Computer Science, vol. 5301, pp. 356–370. Springer, Berlin (2008)
25. Farail, P., Gaufillet, P., Canals, A., Le Camus, C., Sciamma, D., Michel, P., Crégut, X., Pantel, M.: The TOPCASED project: a toolkit in open source for critical aeronautic systems design. Ingénieurs de l'Automobile **781**, 54–59 (2006)
26. Lang, F.: Exp.Open 2.0: a flexible tool integrating partial order, compositional, and on-the-fly verification methods. In: Integrated Formal Methods, pp. 70–88. Springer, Berlin (2005)
27. Garavel, H., Lang, F., Mateescu, R., Serwe, W.: CADP 2010: a toolbox for the construction and analysis of distributed processes. In: Abdulla, P.A., Leino, K.R.M. (eds.) Tools and Algorithms for the Construction and Analysis of Systems. Lecture Notes in Computer Science, vol. 6605, pp. 372–387. Springer, Berlin, Heidelberg, Saarbrücken (2011)

Model-Driven IVV Management with Arcadia and Capella

Jean-Luc Voirin, Stéphane Bonnet, Véronique Normand and Daniel Exertier

Abstract In the field of Model-Based System Engineering (MBSE), this paper describes the use of engineering models to drive and secure Integration Verification Validation (IVV) phases of an engineering lifecycle. The illustration uses the Arcadia engineering method and its supporting modelling workbench Capella. The methodological, tool-agnostic concepts are presented, alongside with tool features easing the application of the approach.

1 Introduction

In the last decade, Model-Based System Engineering (MBSE) has begun taking an increasing place in systems engineering practices, besides a more traditional use of textual requirements. Modeling languages are more and more used today, either for need description (e.g. Architecture Frameworks such as NAF [1]) or for solution description and design (such as SysML language [2]). Many tools exist to support these modeling activities and a few methods are also available (see [3]).

J.-L. Voirin (✉)
Thales Systèmes Aéroportés, 10 Avenue de La 1ère DFL,
Brest 29200, France
e-mail: jean-luc.voirin@fr.thalesgroup.com

S. Bonnet · D. Exertier
Thales Corporate Engineering, 19/21 Avenue Morane Saulnier,
Velizy-Villacoublay 78140, France
e-mail: stephane.bonnet@fr.thalesgroup.com

D. Exertier
e-mail: daniel.exertier@fr.thalesgroup.com

V. Normand
Thales Research and Technology, Route Départementale 128,
Palaiseau 91120, France
e-mail: veronique.normand@fr.thalesgroup.com

G. Auvray et al. (eds.), *Complex Systems Design & Management*,
DOI 10.1007/978-3-319-26109-6_6

However, while most of these methods provide guidance to describe engineering artifacts, very few address the full scope required in real life, large and complex projects: aiding to define and evaluate the architecture; embracing several levels of engineering (system, subsystems, software and hardware engineering); dealing with co-engineering for engineering specialties and analyses such as performance, safety, security, product line, reuse…; aiding and securing Integration Verification Validation (IVV) phases.

This lack of overall methodology led Thales to develop both Arcadia, its own methodological framework addressing all these needs in a comprehensive manner, and Capella, a supporting modeling workbench easing the implementation of the Arcadia method in systems engineering teams.

This paper focuses on one specific engineering concern: how MBSE can ease and secure IVV.

The traditional use of textual requirements is first quickly introduced, along with its perceived limitations in our experience. Then, after a short introduction to Arcadia and Capella, different uses of engineering models in IVV contexts are described: capturing need and solution, building traceability and justification links, aiding in building the IVV strategy, dealing with day-to-day ups and downs. Finally, some ideas for future work or complements in progress are highlighted.

2 Limits of a Sole Requirement-Based Integration, Verification, Validation

One of the major practices today in Systems Engineering consists in relying on textual requirements (often shortened hereafter to ‘requirements’) as the main vehicle for technical management of the contract with the customer, but also as a mean to support need analysis, to define expectations on subsystems, and to drive the Integration Verification Validation (IVV) of the system.

Focusing on the latter, an IVV strategy is usually built during the definition and design phases, in order to define a stepwise integration; the contents of each increment (hereafter called ‘delivery’) is selected, then test scenarios/cases are defined and grouped in test campaigns, as expected to verify requirements for this particular delivery. The verification process mainly relies on the use of traceability links between need definition, solution description and test campaigns/cases.

In the traditional textual requirements-based approach, traceability links are manually created between each requirement and covering test cases. Each delivery is defined as a set of requirements. Then, when all test cases traced towards a requirement are successfully run, this requirement is considered as verified (Fig. 1).

Similarly, other traceability links are built between each requirement and Configuration Items (CI) of the Product Breakdown Structure (PBS)—typically components or subsystems to be integrated—according to the expected contributions of these CI to fulfill the requirement. By this way, the list of CI to be supplied for this IVV phase can then be established.

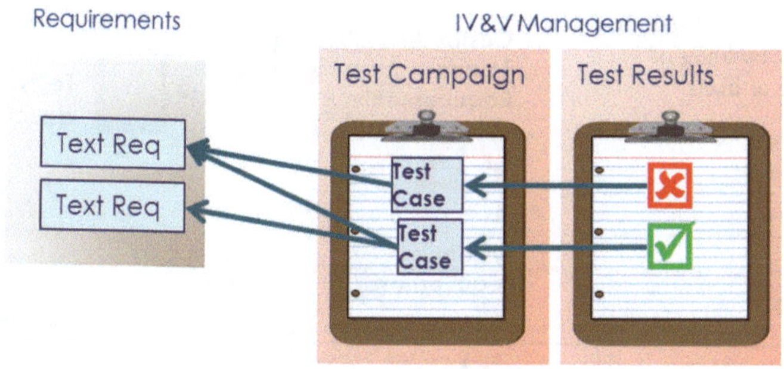

Fig. 1 Traceability links between requirements and tests cases

Experience shows that when the complexity increases, the textual requirements-based approach reaches its limits.

There are many reasons for that:

- Requirements are not able to formally describe the solution and justify it, which means links with PBS are difficult to build and check. The contribution of each CI to a requirement is difficult to identify.
- Traceability links are unreliable and difficult to verify/justify, because their manual building is not formalized.
- The way test campaigns are built remains informal and subject to interpretation or errors.
- In the absence of a precise and detailed vision of architecture and system behavior, it is difficult to clearly identify and localize problems and required changes, and to optimize the testing strategy and non-regression testing accordingly.

The problem mainly lies in trying to create engineering artifacts in a non-formalized manner, with little methodological guidance. This unavoidably generates losses and misunderstandings in the successive transformations/productions of artifacts, as illustrated in Fig. 2.

Consequences on engineering are of particular importance for IVV:

- Requirements are allocated but the solution architecture is often insufficiently described
- Definition of suppliers delivery is weak and not sufficient
- Justification of definition is poor and unreliable
- Checking quality of the definition is not possible before IVV
- Integration verification validation is too expensive due to
  - Wrong estimates and unbalanced incremental IVV
  - Missing components when running planned tests
  - Difficulty to precisely localize/analyze defects
  - Costly and complex non regression testing
  - Poor mastering of changes impact and IVV configurations.

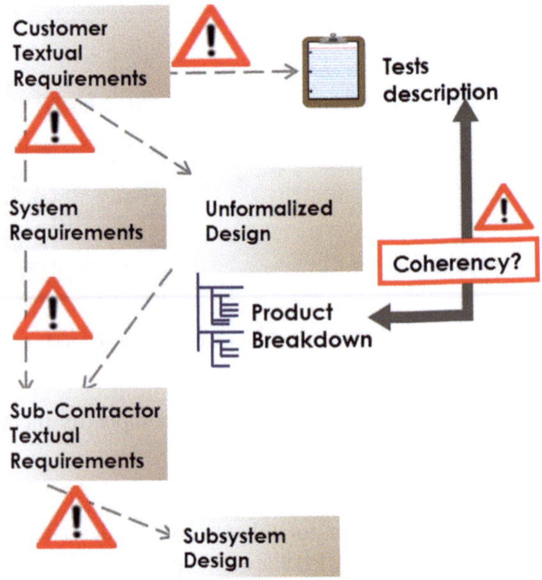

Fig. 2 Non-formalized, thus error-prone, building of engineering artifacts

These problems increase along with system or project complexity, sometimes making IVV become the most costly part of engineering, with a hardly predictable duration.

3 Introducing Arcadia and Capella

Arcadia [4, 5] is a model-based method devoted to systems, software, hardware architecture engineering. It describes the detailed reasoning to understand the real customer need, define and share the product architecture among all engineering stakeholders, early validate its design and justify it, ease and master Integration, Validation, Verification (IVV).

It can be applied to complex systems, equipment, software or hardware architecture definition, especially those dealing with strong constraints to be reconciled (cost, performance, safety, security, reuse, consumption, weight…).

Arcadia is intended to be embraced by most stakeholders in system/product/software/hardware definition, and by IVV actors, as their common engineering reference.

Arcadia has been experimented and validated in many real life contexts for several years now, in most Thales operational units. Its large adoption in many different engineering contexts witnesses of an industry-proven comprehensive method for system engineering, adapting to each context in a dedicated manner, and yet being tooled by the same powerful tools capitalizing know-how.

Some noticeable features of Arcadia are:

- Model-based and tool-supported
- Supporting collaboration and co-engineering
- Open to domain-specific added value
- Adapted to several lifecycles, workshares, etc.
- Dealing with complexity and size
- Field-proven in real industrial situations

As mentioned previously, the field-proven modelling workbench Capella (see [6, 7]) has been developed, both to drive users in applying the Arcadia method and to help them manage complexity with automated simplification mechanisms. A model is built for each Arcadia engineering step. All of these models are related by justification links and are processed as a whole for impact analysis. Arcadia is now partially published and a full publication is on its way. Capella is available as an open source software.

A specific feature of Capella is its ability to support viewpoint-based analysis: the workbench can easily be extended in order to add new specific concepts, link them to the Arcadia concepts, define model analysis rules and display in any regular model diagrams, the effect and results of this analysis.

This approach has been followed in Thales to specify and develop a viewpoint dedicated to IVV management. Screen dumps presented hereafter are extracted from Capella enriched by this proprietary IVV viewpoint. They are not supposed to be readable in details, but rather to be evocative of a high level overview of tool capabilities.

4 Model-Based Traceability/Justification Links Definition

Among other goals, new model-based engineering approaches such as Arcadia aim at overcoming textual requirements limitations.

The need, originally expressed as textual requirements, is formalized in a shareable form that can easily be analyzed and validated:

- An operational analysis describes the operational expectations, goals, activities of the system end-users
- A functional/non-functional analysis, consistent with operational analysis, translates requirements into model elements such as *functions*, *data flows*, etc.
- Traceability links are created between these model elements and originating textual requirements.

Textual requirements are thus complemented and validated, and not replaced, by models. *The need model becomes the major reference for need understanding and impact analysis.*

The solution architecture is formalized, traced, justified and partially validated by an architecture model:

- A functional and non-functional analysis describes the solution expected behavior. This functional analysis results from confronting functional need, non-functional constraints and architecture/design drivers,
- Grouping or segregating the *functions* into *components* (subsystems, software, hardware), under well-defined architectural constraints, leads to the definition and justification, through *data flows* between respective *functions*, of the interfaces between architecture components.

Internally, *models carry most of the description of need and solution*:

- Anything that can be efficiently expressed in the model is formalized that way. One could talk about "modeled requirements". As a consequence, it is unnecessary in that case to create or refine textual requirements, as it would be redundant.
- Internal, textual requirements are added where necessary:
    - To express a constraint or an expectation more precisely than the model,
    - When it is too difficult to represent and capture an expectation in the model
- The customer requirements (UR) remain traced in the model and towards engineering artifacts, for justification purposes.

A significant improvement of engineering becomes possible, as traceability links can now be based on a unifying model. An explicit and verifiable process ensures the quality of those links and therefore, globally secures the engineering (Fig. 3):

- Justification links between requirements and *functions* are created when translating textual requirements into model elements,
- Realization and allocation links between *functions* and *components* result in PBS definition/justification/traceability links,
- IVV tests are based on model elements and not anymore on possibly ambiguous textual requirements,
- Impact analysis can be performed across system to subsystems models.

Links between requirements and IVV tests are derived from requirement-to-model and model-to-tests links, which makes them more reliable and easier to review and check.

Verification of textual requirements is ensured through the model, exploiting the tests-to-model elements links, and the model-to-requirements links:

- A model element is considered as verified when all linked tests are passed
- A requirement is verified when all linked model elements are verified.

Note that one additional benefit of the model is also the enrichment and reinforcement of the definition/technical contracts of subsystems and components.

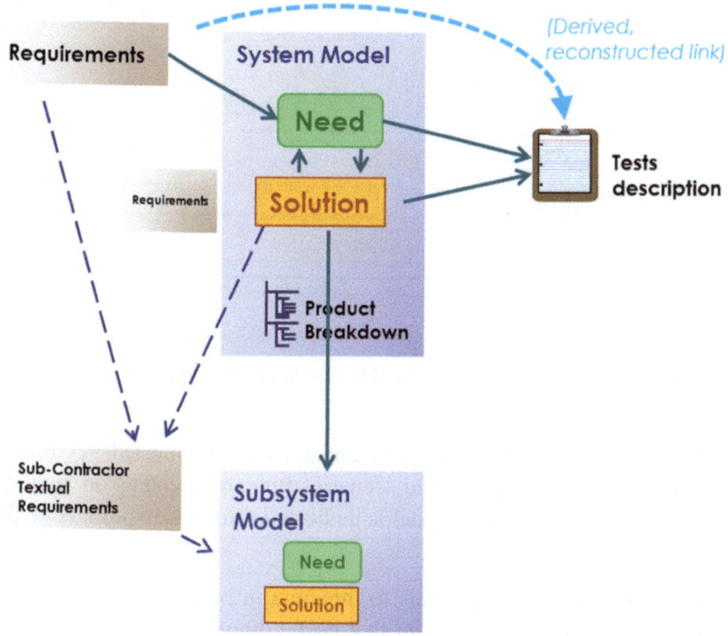

Fig. 3 Model-driven engineering and traceability

5 Building an IVV Strategy

Building IVV strategy mainly deals with defining test and integration increments, in order to progressively verify the system solution and its adequacy to the need.

In the Arcadia method, a 'requested version' (RV) of the system is defined for each IVV milestone. The content of a RV is not described by a set of textual requirements, but rather by a set of model artifacts such as *capabilities* (use cases), *scenarios* (sequence diagrams) and *functional chains* (ordered set of functions and *data flows*), expressing the dynamic use of the system, as specified in operational and system need analysis, and as designed in solution architecture model. A first benefit comes from defining versions contents, not as abstract requirement sets, but as *customer-friendly capabilities and use cases*.

These artifacts are by nature related to other model elements such as *functions*, *data flows*, *components*, *data*, *interfaces*, non-functional properties, etc. They also are directly used to define test campaigns, test suites and test cases, to which they are linked by traceability/justification links (Fig. 4).

Test campaigns are constructed from model *scenarios* and *functional chains*, refining and detailing them (limits and nominal testing, non-nominal cases...).

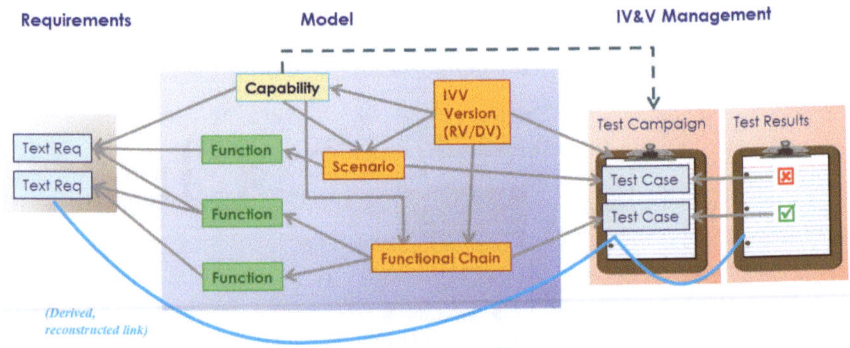

Fig. 4 Model-driven IVV strategy building (DV notion described later)

Traceability links are built between test campaigns and Requested Versions (RV) on one side, and between test cases and scenarios/functional chains on another side.

Based on the analysis of the model, a dedicated tooled viewpoint (Fig. 5) is then able to automatically:

- Determine the list of components to integrate in order to feed this delivery,
- Summarize the functional content that each component needs to provide,
- Specify the expected testing means functional content. This allows an incremental development and delivery of test means, which significantly de-stresses their engineering.

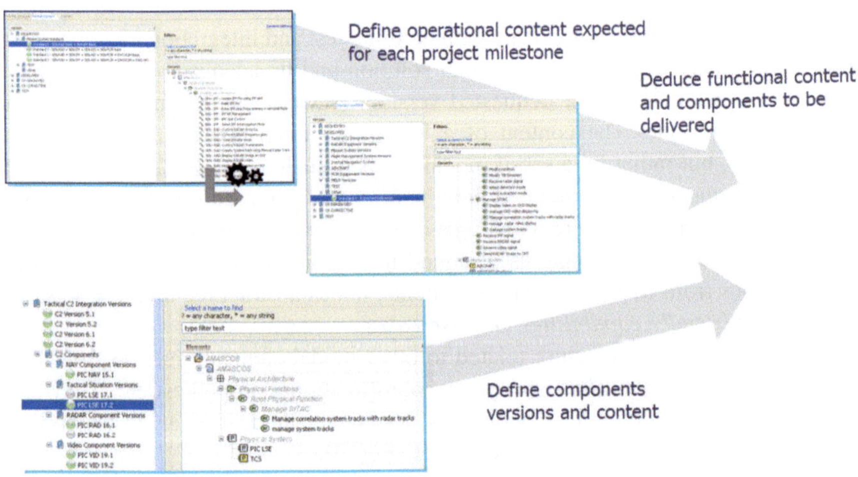

Fig. 5 Tooled process to create versions contents

6 Day to Day IVV Activities Model-Based Support

In the daily life of the IVV manager, the first benefits of a model-driven IVV come from a much better mastering of the system architecture, a better understanding of its behavior and greater fault localization accuracy.

But management of unavoidable ups, downs and hazards occurring during IVV is also greatly facilitated. For this purpose, the notion of actually 'Developed Version' (DV) has been introduced, so as to capture the real structural and functional contents of each built version, according to the real state of components delivered at this moment. This allows analyzing what is really available due to components actual contents, and evaluating how the initial integration strategy has to be adapted accordingly (Fig. 6).

For example, if a component or subsystem is delivered late, or if its functional content does not conform to what was planned in the requested version, the use of the model and associated viewpoint tooling allows to identify the operational capabilities or features that are consequently not available.

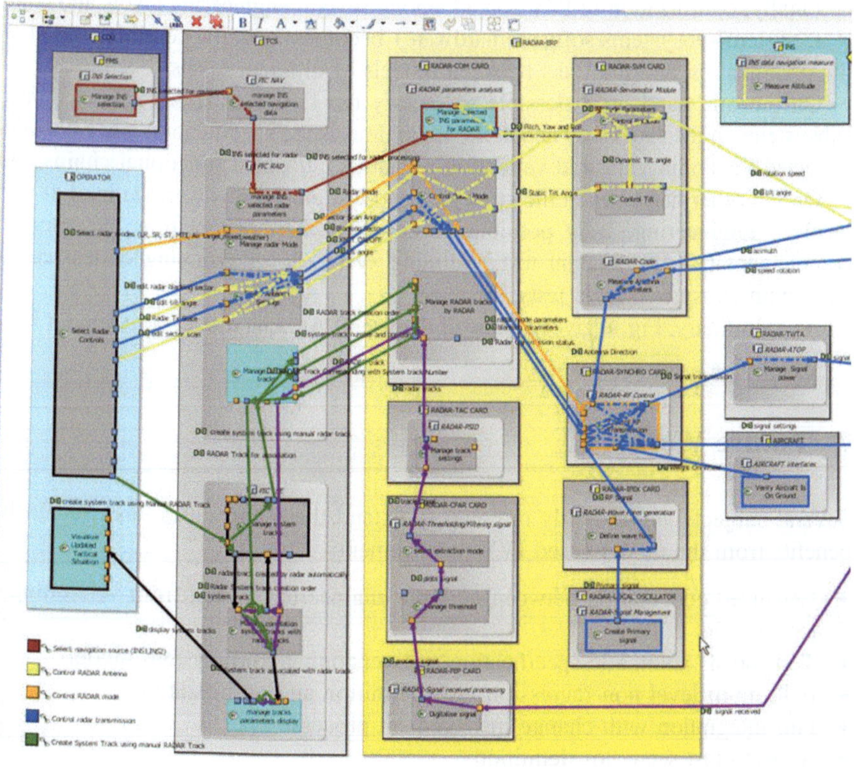

Fig. 6 Decoration of any diagram with delivered versions contents and impact analysis results: in *grey*, what was expected and is available at integration time; in *cyan*, what was expected and not available; colored paths are functional chains to be run for testing this version—if possible

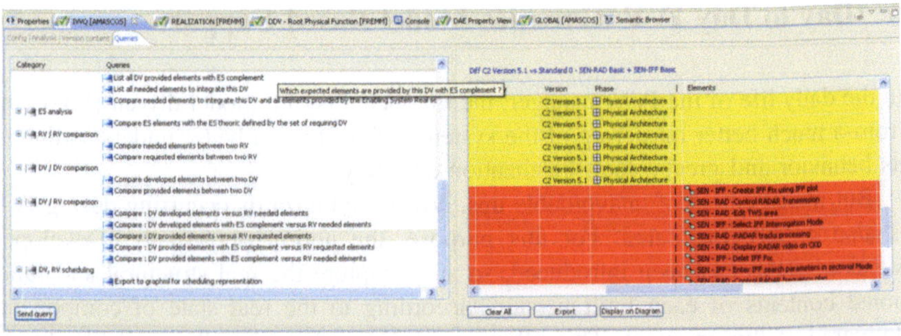

Fig. 7 Impact analysis rules (here scenarios and functional chains that cannot be run due to incomplete components contents)

Similarly, the tool can identify the tests, or *functional chains*, which are not to be run because of missing or incomplete *components*. The regression tests can also be optimized based on the functional content that has changed between two versions or after corrections of defects (Fig. 7).

Finally, optimization of IVV spread among several engineering levels and teams (e.g. system/subsystems/software/hardware) becomes actually possible, based on the organization and the links between models of different levels.

On the one hand, it is possible to specify the IVV campaigns expected from subsystems in the same way and at the same time as the specification laid by model: expected versioning, desired validation scenarios, allocated functional chains, etc.

On the other hand, when the strategies and test campaigns were defined at each level of engineering, it is possible to optimize their articulation by detecting complementary or redundant tests, defining system-level grouping of test coherently with subsystem-level tests, etc.

As a summary, Fig. 8 gives a view of the use of Arcadia model contents for IVV.

7 Future Work

Several subjects are currently ongoing to extend these capacities, and get further benefits from the model-based IVV management:

- Enhanced cooperation between multiple engineering levels, from versioning to tests results
- Test means automatic specification and versioning from system model
- Full, multi-level non regression tests definition and optimization
- Full integration with change management process
- Assisted test scenarios definition
- …

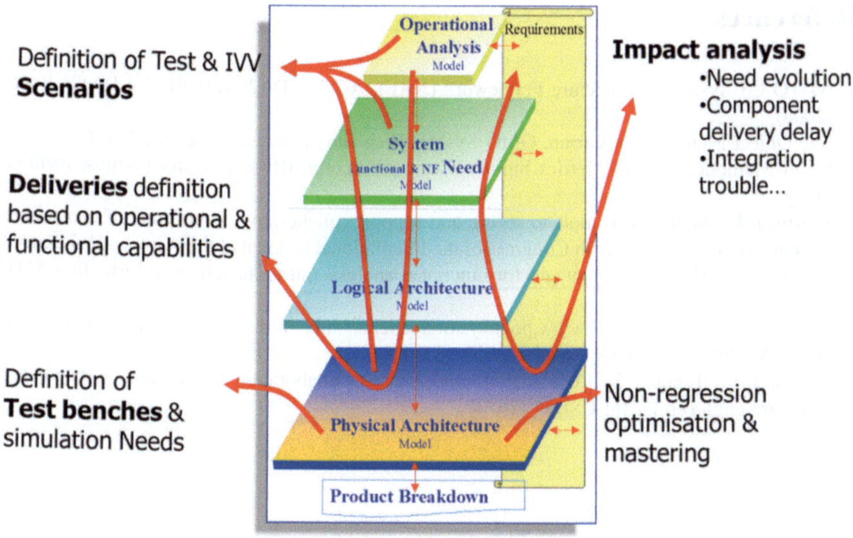

Fig. 8 Summary of model uses for IVV (*arrows* show which part of the model contributes to each IVV task)

8 Conclusion

This paper has introduced an innovative model-based management of Integration Verification Validation (IVV) processes, based on some major orientations:

- Moving from requirement-driven integration
- towards model-based, *scenarios* and *functional chains*-driven IVV
- Anticipating the functional impact of delayed or missing *functions/components* at integration time
- Analyzing the impact of *component* change or lack of maturity, and adjusting the testing strategy accordingly.

The objective of this paper was to provide a first sight on the major principles of this model-driven IVV approach. As a result of first operational uses of the approach, models appear to be a powerful lever to improve IVV practices:

- Build, validate, optimize and master IVV strategy
- Understand and check system behavior and architecture
- Build balanced IVV versions, securing operational contents
- Describe and plan tests, securing the ability to run them
- Specify test means, securing their adequacy to each IVV version
- Manage impact of Change Requests (CR)
- Manage unexpected events (late component delivery, CR pending, etc.)
- Use shareable, formalized models for that purpose,
- Ensure automatic impact analysis and check capabilities.

References

1. NATO C3 System Architecture Framework (NAF), AC/322-D (2004)0041, NATO C3 Board (2004)
2. The Object Management Group, OMG SysML Web site http://www.omgsysml.org/
3. OMG Methodology and Metrics http://www.omgwiki.org/MBSE/doku.php?id=mbse:methodology
4. Voirin, J.-L.: Method and tools to secure and support collaborative architecting of constrained systems. In: ICAS 2010, 27th Congress of the International Council of the Aeronautical Science
5. Voirin, J.-L.: Modelling languages for functional analysis put to the test of real life. In: CSDM (2012)
6. Capella web site, https://www.polarsys.org/capella, and https://www.polarsys.org/capella/publis/An_Introduction_to_Arcadia_20150115.pdf
7. Voirin, J.-L., Bonnet, S.: ARCADIA: model-based collaboration for system, software and hardware engineering. In: CSDM (2013)

How to Make Sure the System Level Conformity Assessment: Case of Japanese Consortia in Automotive Communication Protocol

Akio Tokuda

Abstract Standards developing organizations (SDOs) have recently emphasized the importance of the system level conformity assessment (SLCA) which is free from the interoperability problems of complex product systems. The objective of this paper is to show how to establish the dependable SLCA and what kind of co-operation among the actors is needed to the establishment of the SLCA. For this purpose, I am dealing with, as the case study, the standard setting process of the conformance test specifications of the automotive network system in Japanese industrial consortium. The case reveals that if a society demands a highly dependable SLCA, vertical co-operation among actors should be designed for drafting the conformance test specifications while horizontal co-operation should be oriented for restricting the universality of the specifications among actors.

Keywords System level conformity assessment · Standard · Interoperability problem · Conformance test · Consortium

1 Introduction

SDOs have recently emphasized the importance of the system level conformity assessment (SLCA) which is free from the system level interoperability problems of complex product systems. As the multiplicity of technologies and their convergence demand a top-down approach to standardization, SDO, such as IEC, has tried to make an improvement of its standardized conformity assessment system from product-oriented to system-oriented one by means of increasing co-operation with many other SDOs (e.g. ISO, ITU) and industrial consortia [1]. However, they have not gained the clear picture on how to cope with the system level interoperability problems.

A. Tokuda (✉)
Ritsumeikan University, 2-150 Iwakura-cho Ibaraki-city, Osaka 567-8570, Japan
e-mail: att20023@ba.ritsumei.ac.jp

G. Auvray et al. (eds.), *Complex Systems Design & Management*,
DOI 10.1007/978-3-319-26109-6_7

The objective of this paper is to show how to establish the dependable SLCA, which is free from the system level interoperability problems, and what kind of co-operation is needed to the SLCA. For this purpose, I am dealing with, as the case study, the standard setting process of the conformance test specifications of the automotive network system at Japanese industrial consortium.

This paper proceeds as follows. The history of standardization of automotive network protocols will be briefly traced and then overview the standardization process of FlexRay, one of the de facto standards of the next generation network protocol in Sect. 2. In Sect. 3, a comparative analysis of the drafting process of FlexRay's conformance test specifications between European and Japanese consortiums will be made. In Sect. 4, the standardization process of Japanese consortium is described more in detail by focusing on the co-operation mechanisms inside it. Finally, I will draw some conclusions from the case study.

2 Standardization of Protocols

2.1 The Distributed Cooperative Control

With the aim of creating new applications such as environment-friendliness and advanced safety, car manufactures are confronted by the need to integrate an increasing number of Electronic Control Units (ECUs) into a single network. For instance, development of the environment-friendly automobile, integration of the engine control unit with the braking control unit and the motor control unit are essential. We are moving to the phase of distributed cooperative control of multiple ECUs to realize such new applications [2, 3].

In order to realize the distributed cooperative control of ECUs, it has been important for car manufactures to standardize network protocols. Today, there are a number of sets of protocols, which serve as international de facto standards in the automotive sector. Figure 1 shows the recent development of standardized protocols by the domains.

In each domain, electronic devices (e.g. sensors, ECUs, and actuators) are connected by standardized protocol. After much conflict between local protocols, several consortiums worked to set standardized domains: a body control domain by Local Interconnect Network (LIN) Consortium, a multimedia control domain by Media Oriented Systems Transport (MOST) Cooperation, a safety control domain by Safe-by-Wire Plus Consortium and a powertrain and a chassis control domain by FlexRay Consortium.

Amazingly, every standardized protocol except Safe-by-Wire was developed via European-origin consortiums. Also these domains are mainly networked via core protocol Controller Area Network (CAN), which was developed by the German system supplier Bosch allied with Daimler. The CAN was the first protocol to gain the de facto standard status in the automotive industry [4].

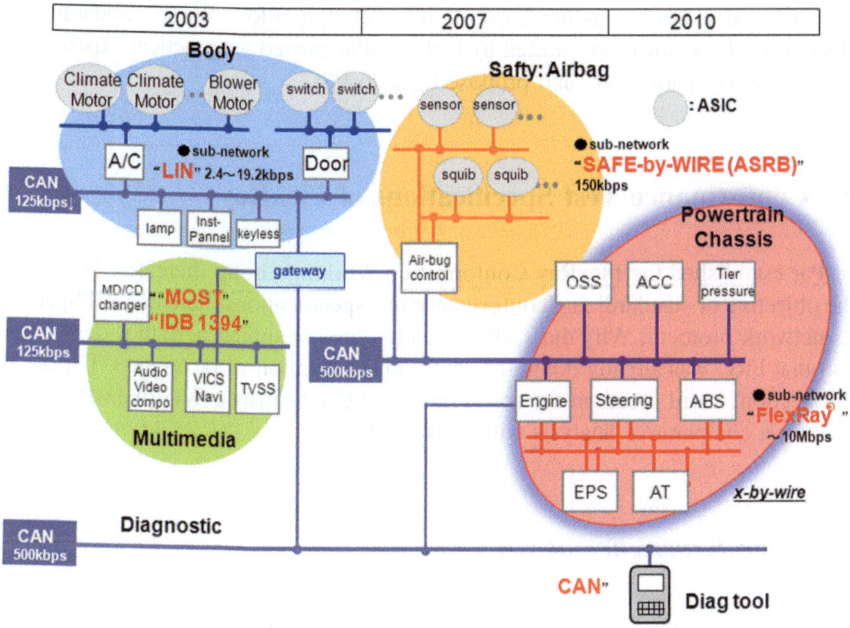

Fig. 1 Standardized protocol by domains (*Source* Renesas Electronics Co., Ltd.)

2.2 Standardization of FlexRay

As a successor protocol to CAN, FlexRay has attracted attention in the automotive sector. In recent years, the volume of data carried by networks has grown hugely, so a protocol with higher communication speed than CAN has become necessary.

In the development and standardization of FlexRay, a standard consortium of European origin, fulfills the leadership role. In 2000, a group of four companies: BMW, Daimler-Chrysler (now Daimler), the semiconductor vendor Motorola (now Freescale) and Phillips (now NXP) formed the FlexRay Consortium (hereafter FRC). Subsequently, Bosch, GM, and Volkswagen joined FRC, and were followed by the Japanese companies Toyota, Nissan, Honda, and Denso [3].

The aim of FRC was to establish a de facto standard by jointly developing a protocol and corresponding systems. While promoting FlexRay with the Society of Automotive Engineers (SAE) and marketing it to US car manufactures, FRC built up a collaborative framework with a corresponding standard setting consortiums such as the Japanese embedded system standardization consortium Japan Automotive Software Platform and Architecture (JasPar), for promoting the diffusion of FlexRay. JasPar is a consortium of Japanese origin established in 2004. Its main objective is to change the product architecture of automotive embedded systems from a vertically integrated architecture to an open disintegrated one by means of modularization of systems with standardized interfaces via collaboration with European corresponding consortiums including FRC. While validating the standardized specifications drafted by FRC, JasPar carried out practical experiments to set concrete parameters and propose them to FRC.

3 Conformance Test Specifications of FlexRay

JasPar established the FlexRay Conformance Working Group (hereafter WG) with the objective of standardizing conformance test specifications (hereafter CTSpec) of its network protocol. Why did JasPar need to conduct similar activities despite the fact that FRC was already doing the same standardization work on the CTSpec? In this section, I will focus on the activities of the Conformance WG of JasPar while making a comparative analysis with those of FRC.

3.1 The Experience of CAN

A conformance test basically means looking to see whether the product has been turned out in accordance with the standardized specifications. The CTSpec of FlexRay is a guideline available to any of FRC's semiconductor vendors when they want to assess whether their microcomputers (hardware) and device drivers (software) conform to the standardized protocol specifications.

Figure 2 shows the flow of the conformance procedure. Based on the CTSpec set by FRC, Conformity Assessment Body (CAB), which is certified by the Accreditation Body (AB: e.g. FRC), assesses devices of semiconductor vendors for certification. Then, system suppliers procure the certified devices from the semiconductor vendors. Finally these certified devices are embedded into ECUs and delivered to car manufactures.

The CTSpec is no more than a test for assessing whether the vendor's devices conform to the specifications on a "stand-alone basis". For Japanese car manufactures, however, it gives the priority to create more system-oriented test specifications which ensure the "system-level interoperability"[1] of these devices. When CAN was introduced to the Japanese market, interoperability problems

[1]In the first instance, two complementary technical components Y and Z are compatible, it is said those components are compatible complements or "vertically compatible". In the second instance not only component Y but also another functionally equivalent component X is compatible with component Z, we can say that with respect to interoperability with component Z, both components are compatible substitute or "horizontally compatible" [5]. In this paper, if the vertically incompatibility happens, it means the situation that there is "component-level interoperability problem (between Y and Z)" while if horizontally incompatibility happens, it means the situation that there is "system-level interoperability problem (between X, Y and Z)". Later type of relationship between components represents more general form of compatibility than that of former.

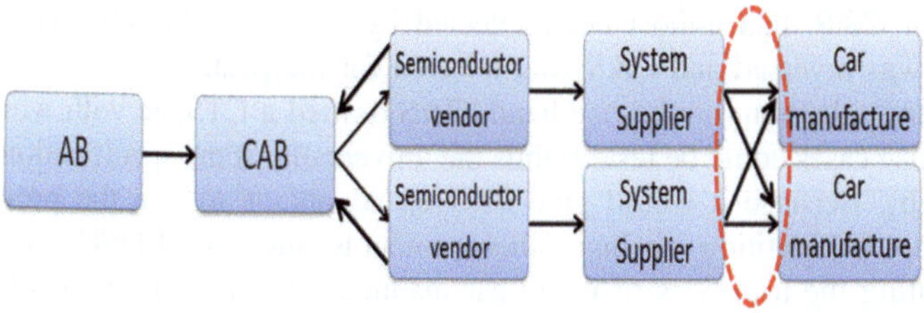

Fig. 2 Procedure of conformance test

frequently occurred even though CAN was supposed to be used as the standardized protocol. Car manufactures were able to connect different ECUs procured from the same supplier, but ECUs from different suppliers weren't interoperable (see Fig. 2 dotted oval).

The reason behind the interoperability problems was, in part, the "abstract" nature of the description of the protocol specifications. This "abstract" nature meant exact parameter values and ranges were not clearly fixed in the specifications, so that the semiconductor vendors had been, to a certain extent, able to interpret them freely. This resulted in discrepancies of devices among vendors despite the fact that they conformed to the standardized specifications.

The second reason the interoperability problem arose was the "imperfect" test coverage of CTSpec against protocol specifications. The semiconductor vendors had utilized CTSpec to interpolate the "abstract" nature of the protocol specifications. However, the CTSpec of CAN does not cover all the tests against its protocol specifications. The validity is determined on the basis of some sampling representative values. If a certain number of sampling values are acceptable, it assumes all values to be acceptable. Thus the vendors had no choice but to add their own test specifications in order to make sure all the values are right. As we will see in the next section, similar concerns regarding system-level interoperability have been raised in the case of FlexRay.

3.2 *The Feature of CTSpec of FlexRay*

The setting process of the CTSpec at FRC marks a sharp contrast with that of JasPar. Despite FRC starting the Conformance WG, the actual development has not been made at the WG, but outsourced to one of the members of FRC, C&S. The only work to be done at the WG was to review the draft developed by C&S, and then nominate conformity assessment bodies who conduct conformance test according to that draft.

The reason why C&S was selected at FRC was largely due to the influence of the head of the WG at that time over the selection process. Other firms such as TTAutomotive and TÜV NORD, which represented a developer and company other than C&S, had almost been selected by a vote at the WG. However, the decision was reversed and C&S was selected for the post.

After the selection, on the one hand, C&S drafted a CTSpec with which a wide range of use cases could be tested (thus the universality if the specification supposes to be high). Because it would give FRC an advantage to win the protocol standardization competition against other protocols such as TTP/C by positively incorporating the use cases of many car manufactures into the protocol specifications. On the other hand, the development and experiment of test cases against a wide range of use cases would incur a large expense. Therefore, C&S made the decision to develop an economy-oriented CTSpec by means of loosening the test coverage. With European conformance test specifications, validity is determined on the basis of a sampling of values. The approach is to test a certain number of a sampling of values and if they are acceptable to assume all values to be acceptable. However, this approach means that no one knows whether all the values are correct.

The method of drafting the CTSpec at FRC reminded Japanese car manufactures of the system-level interoperability problems, which happened during the introduction stage of CAN that caused them to decide to develop the improved version of a FlexRay CTSpec by themselves.

3.3 The CTSpec Setting Process at JasPar

In order to develop a more dependable CTSpec, the horizontal and the vertical co-operation between working groups have been enhanced at JasPar. Firstly, the horizontal co-operation among car manufacturers attempts to narrow down their use cases at the cost of universality of the CTSpec. Then, vertical co-operation enhances the use cases to identify vertical interoperability problems between complementary components. Finally horizontal co-operation is made via collaborative experiment and assessment among semiconductor vendors for securing horizontal compatibility of each alternative device. Based on this co-operation, test items with 100 % coverage against narrow use cases were created at JasPar.

To give a specific instance, paying attention to 97 pieces of Service Description Language (SDL) charts in which all of behaviors of microcontrollers are described, JasPar tested path checks against all the charts and then drafted specifications that give 100 % path coverage. JasPar has likewise added test items to their CTSpec, which are not covered by those of FRC. After verifying the FRC's physical layer specifications relevant to the bus driver, JasPar identified FRC covered only 140 test items against the FRC's physical layer specifications. As a result, they added 66 test items to it, so that the test coverage reached 100 % (Table 1).

Figure 3 describes the conceptualized characteristics of the CTSpec of each consortium. FRC standardized their CTSpec, which covers the full range of use cases, while it does not cover the full range of test items against its standardized specifications. By contrast with FRC, JasPar narrowed down the use cases keeping its test coverage at 100 %.

Table 1 The number of test items of each consortium

	FRC's physical layer specifications	FRC's CTSpec against its PLSpec	JasPar's addition for FRC's CTSpec
The number of test items	206	140	66

Source JasPar

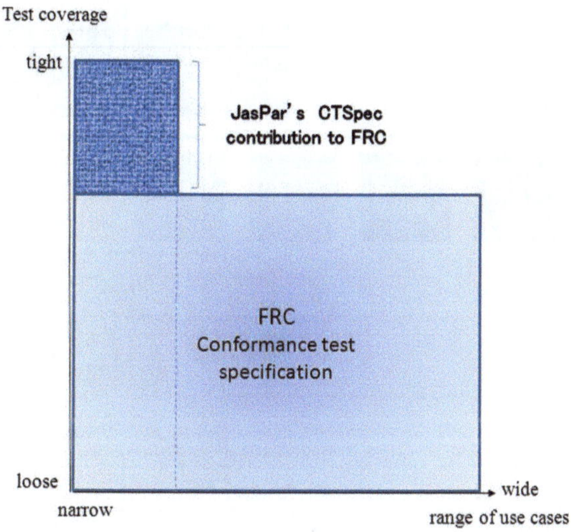

Fig. 3 Relationship of CTSpec

We could paraphrase this relationship as saying that FRC tried to standardize a relatively "wide and loose" CTSpec while JasPar tried to develop a "narrow and tight" one. Both types of CTSpec have their pros and cons, but my research is to clarify what brought them to this difference in drafting their CTSpec, by focusing on the co-operation mechanism of JasPar in the following section.

4 The Co-operation Mechanism at JasPar

When it comes to the setting of a standardized interface, which is useful for ensuring the system-level interoperability, there may be a limit to what individual suppliers can achieve alone. In order to make sure of system-level interoperability, vertical and horizontal co-operation is called for among the concerned car manufacturers and suppliers. How has such co-operation been conducted at JasPar? I will briefly introduce the co-operation mechanism of the Automotive LAN WG at JasPar.

Figure 4 shows the organization chart of JasPar in 2008. Since its foundation, JasPar has devoted energies mainly to the activity of the Automotive LAN WG. The ultimate mission of the WG, together with its subsidiary WGs and task forces, is to draft the "narrow and tight" CTSpec of FlexRay.

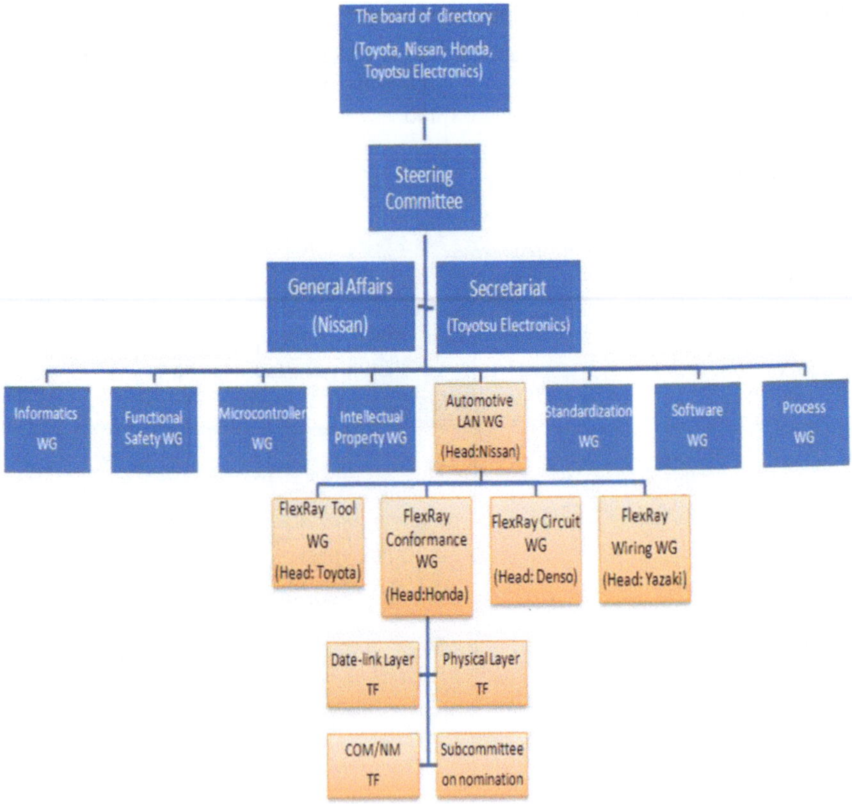

Fig. 4 Organization chart in 2008 (*Source* JasPar)

The Automotive LAN WG itself has been operated by car manufactures. One of the most important tasks of the WG was, via horizontal co-operation among car manufactures, to narrow down the use cases in order to reduce the workload of the relevant subsidiary WGs while making the setting of standardized default values of parameters quite easy. In addition, the two following types of co-operation were carried out under the WG.

4.1 The Vertical Co-operation for Component-Level Interoperability

With recent high-speed networks, wiring has to be designed with a deep understanding of physical conditions on the circuit side, due to interference interdependence across components which cause component-level interoperability problems (e.g. physical, electrical and electro-magnetic noise interfere sampling at bus drivers). The Circuit WG at JasPar carried out a series of experiments to gather information regarding the cause of the interoperability problems and used them as the basis for developing the recommended circuit. The Wiring WG works alongside this activity, identifying the characteristics required in the bus driver from the wiring simulation data and presenting this data to the Circuit WG. It is after this continuous circulation of information that the Circuit WG finally completes the recommended circuit (Fig. 5). The Wiring WG uses the circuit as the basis to carry out the final wiring simulation. In this way, the collaboration between the circuit side and the wiring side is maintained.

To put it another way, JasPar shares information via vertical collaboration between WGs, then tries to find latent component-level interoperability problems across the WGs, and finally finds solutions to them for setting an interface, which is free from any intervention between complementary components.

4.2 *The Horizontal Co-operation for System-Level Interoperability*

Through vertical collaboration with the Wiring WG, the Circuit WG carries out experiments repeatedly in connection with different ECUs in order to make sure of their horizontal compatibility (Fig. 6).

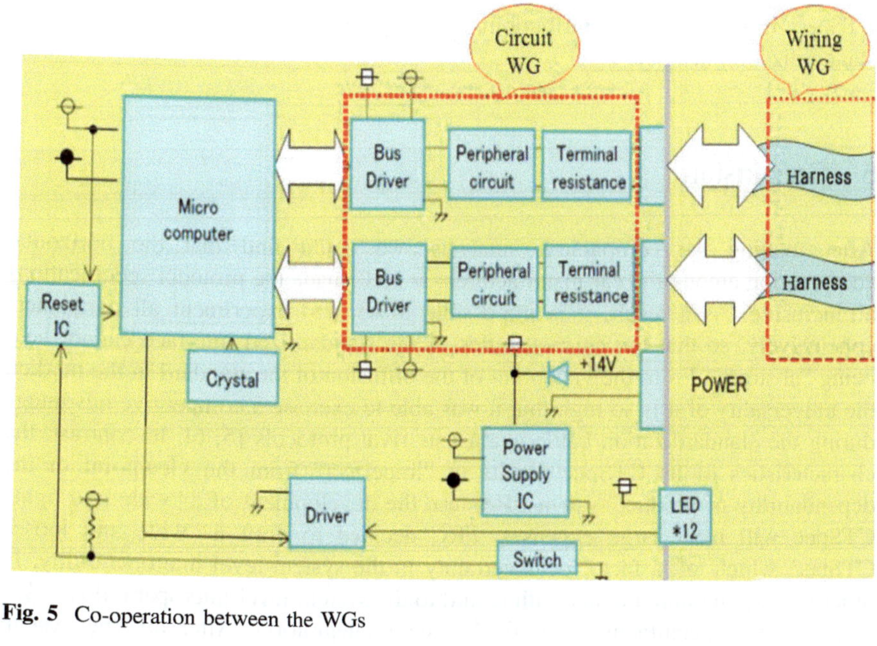

Fig. 5 Co-operation between the WGs

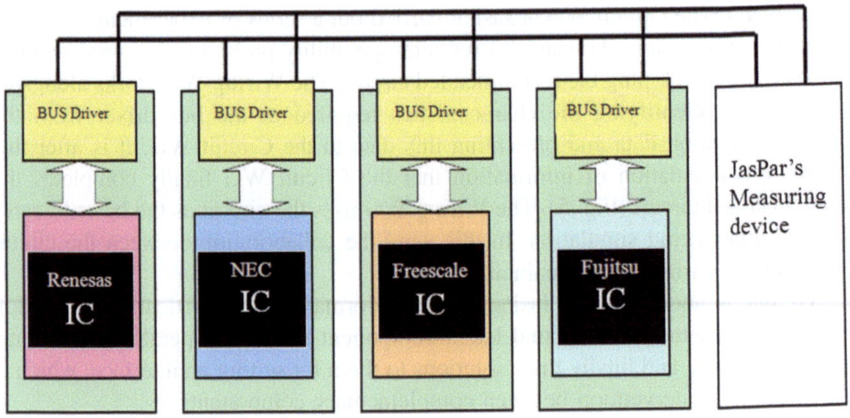

Fig. 6 The horizontal interoperability test

With this experiment, the Circuit WG can identify system-level interoperability problems between standardized bus drivers and alternative semiconductors' ICs, which could be the cause of the horizontal incompatibility between them. Then information relevant to system-level interoperability problems accumulated at the WG will be utilized at the Conformance WG for drafting the tight CTSpec.

It is appropriate to draw conclusions that such vertical and horizontal co-operation among WGs may be unique to JasPar and allowed them to produce the essential input into the Conformance WG for drafting the "narrow and tight" CTSpec. It is one of the contributions by Japanese consortium to Europe-centered standardization activity.

5 Conclusion

After making the comparative analysis, we could find that the horizontal co-operation among the car manufacturers at FRC made the protocol specifications all-inclusive. As a result, it is impossible to set and experiment all parameters conclusively, so that the characteristics of the standardized interface cannot help being "abstract". From the viewpoint of the diffusion of the standard in the market, the universality of it is so high that it was able to exercise a competitive advantage during the standardization battle against its rival protocols [5, 6]. In contrast, the characteristics of its CTSpec results in "imperfect" from the viewpoint of the dependability of product systems. Because the development of a "wide and tight" CTSpec will incur large expenses, FRC decided to draft a "wide and loose" CTSpec, which left a factor of uncertainty in the system-level interoperability. In other words, premium quality with regard to the system-level interoperability is left to automotive manufactures (or final system integrators). After all, the system integration capability of the firms still ought to be counted even in the world of modularization with standardized interfaces.

On the other hand, JasPar attempted to narrow down use cases through horizontal co-operation among car manufacturers. Also, vertical co-operation has been enhanced to identify both logical and physical vertical incompatibility between complementary components. Lastly, horizontal co-operation has been made among semiconductor vendors for securing system-level interoperability. Based on these co-operations, a "wide and tight" CTSpec, which is expected to be useful for securing system-level interoperability of the product system, is created at JasPar. To see it from another angle, the quality premium of system-level interoperability is embedded into the standardized interface ex ante.

Under a clear strategy not to make the same mistake as CAN did, JasPar tried to develop the CTSpec free from system level interoperability problems via both vertical and horizontal co-operations inside the consortium. We can enjoy a variety of benefits that come from the world of modularization with standardized interfaces only if we can successfully set the standards dependably through adequate ex ante co-operation among firms.

I hope any SDOs which emphasize the importance of the system level conformity assessment of complex product systems (e.g. Cyber Physical System) would learn some lessons from the case study, and I suppose one of the important roles of official SDOs is, as socio-technological entities, to keep a balance between the dependability and the universality of the standardized interface of the systems by designing the necessary forms of co-operations among the actors, those are suggested in this paper.

References

1. International Electrotechnical Commission.: In: IEC Master Plan 2011, IEC (2011)
2. Gerybadze, A., König, R.: Managing global innovation networks: the case of automotive electronics. In: Presentation at Workshop on Managing Global Innovation Networks, Duisburg University, 8 Feb (2008)
3. Murray, C. J.: Four Asian automaker join FlexRay consortium. Electron Eng Times, 1 March (2004)
4. Tokuda, A.: International framework for collaboration between European and Japanese standard consortia, pp. 157–170. In: Jacobs, K. (ed.) Information and Communication Technology Standardization for E-business Sectors. IDEA Group Publishing (2009)
5. Katz, M.L., Shapiro, C.: Systems competition and network effects. J Econ Perspect **8**(2), 93–115 (1994)
6. Gabel, H.L.: Competitive Strategies for Product Standards. McGraw-Hill, London (1991)
7. Schmidt, S.K., Werle, R.: Coordinating Technology: Studies in the International Standardization of Telecommunications. The MIT Press, Cambridge, Massachusetts, London England (1998)

Analysis of Implementation of Care Coordination in a Multi-level Care Provider Organization: A Need for Systems Approaches

Guillaume Lamé, Tu-Anh Duong, Marija Jankovic, Julie Stal-Le Cardinal and Oualid Jouini

Abstract Better care coordination is a crucial objective to answer to the rising complexity of healthcare and the associated increase in costs. Process-based organizations is a widely recommended method for achieving this goal. In this article an initiative of implementing a care process in a French public hospital group is analyzed. The procedure to design the care process is documented and the official care process is compared to the current situation in a hospital. This analysis shows how important local parameters are in such projects. The shortcomings of the approach are identified and propositions to overcome these issues are made.

1 Introduction

Health care systems around the world are under pressure. As life expectancy is increasing and chronic diseases are getting more frequent, demand is rising but national expenditure is not rising as fast [1]. Therefore there is a clear need for productivity increase. In France, hospitals have been identified as one of the sources of increase of global efficiency [2].

However, in the same time, care complexity is continuously increasing [3], and hospitals have long-since been identified as complex organizations [4]. In this ever-increasing complexity, sources of productivity need to be identified and efficiency losses need to be addressed.

To deal with this situation, organ-based silo organizations are no longer adequate and better care coordination should be attained [5]. Systems engineering approaches have been identified as a way forward [6], and as a part of this process-based

G. Lamé (✉) · T.-A. Duong · M. Jankovic · J.S.-L. Cardinal · O. Jouini
Laboratoire Genie Industriel, Ecole Centrale Paris, Grande Voie des Vignes, Chatenay-Malabry 92290, France
e-mail: Guillaume.lame@ecp.fr

G. Auvray et al. (eds.), *Complex Systems Design & Management*,
DOI 10.1007/978-3-319-26109-6_8

organizations have been promoted to achieve better coordination. Consequently, hospitals have started to shift towards more process-oriented organizations by designing care processes. However, social scientists have shown that theoretical organizations and real-life events are not the same thing [7]. More particularly Nyssen [8] shows that real-time care coordination can be an "emergence-through-use" phenomenon. The intent of the designer of the management procedures and the behavior of the real system can be two very different things.

In this context, the objective is to achieve coordination by design in order to have more robust and predictable coordination mechanisms. How can this be achieved? To answer this question, an initiative from a French care provider is studied in this article. It is a top-down project aiming at the implementation of an integrated care process for cancer care. The coordination mechanism "as designed" is compared to the situation "as happening". Dates of main events are gathered from medical records to build a picture of the real current situation. This situation is confronted with the desired model.

Conclusions are drawn for the design of coordination-enabling systems and processes and the management of large complex organizational systems. Centralized and non-differentiated initiatives have poor chances of success. In this case, taking into account the diversity of cancer types and organizations is crucial.

2 Context and Literature Review: Care Coordination

2.1 Care Coordination

Care coordination has been defined as "*the deliberate organization of patient care activities between two or more participants (including the patient) involved in a patient's care to facilitate the appropriate delivery of health care services.*" [9] It includes resource management and is often achieved through information exchanges. To achieve this, process-oriented care organizations have been proposed as a solution [10]. Different approaches exist to implement process-based care organizations, which can be classified in two broad categories [11]: an "industrial" approach based on methods such as lean management, total quality management and business process reengineering [12]; and the "integrated care" stream, coming from the medical world, with a strong emphasis on evidence-based medicine [11, 13]. The organization studied here applied the "integrated care" approach, although evidence is limited on the efficiency of such a method (Vanhaecht shows that integrated care pathways improve some aspects of care coordination [13], but other researchers exhibit mixed results for clinical pathways initiatives [14, 15]).

2.2 French Context for Cancer Care Coordination

In France, cancer care coordination is organized as a multi-level system. Table 1 shows the different levels of care coordination at the national and regional levels and for the case of the care provider and the hospital at study.

This provider is a public grouping of university hospitals, hospitals and clinics. It has elaborated a cancer strategy to comply with national and regional directives and to maintain its leadership on cancer care and research. This strategy is informed by a cancer working group at the head of the organization. Expert centers are certified in the member-hospitals of the provider. This certification is an acknowledgement of their excellence for care and research.

As part of its cancer strategy, the provider has launched an initiative to improve its organization for providing cancer care. This initiative is related to the "integrated care pathways" concept. A care process has been designed for implementation in all member hospitals, and indicators have been defined to manage this process.

In this study the objective is to analyze this project and contrast it with systems engineering perspectives. Indeed, previous studies have shown limited adherence to clinical pathways [16] and mixed results when comparing theoretical processes with real-life events [7, 8], which raises questions on the approach used.

Usually, systems engineering starts with the analysis of context, environment, shareholders and objectives. Then specifications are established for the system. Finally processes are designed [17]. This phase is top-down, from the system towards its elements. The design of the care process is documented and situation at the operational level, i.e. the "down" part of this top-down initiative, is analyzed to draw conclusions on the approach used.

Table 1 Levels of cancer care coordination in France

Level	Actions and responsibilities	Figures
France	Plan Cancer 2014–19: strategy and objectives National Cancer Institute (INCa): recommendations and certification	3 million patients 355.000 new cases/year 148.000 deaths/year
Regional health agency	Transcription of national directives into regional policy Audit	
Provider	Cancer working group 37 Expert Centers	30 % of region's cases 83.500 patients/year
Hospital	3 Expert Centers 1 Cancer Coordination Center	

3 Methods

First, a qualitative study is conducted to understand how the process for cancer care was designed. This analysis is performed by document analysis and a 45 min semi-directed interview with the head of the oncology department at the studied hospital, who was also a member of the process design project team.

Then data is analyzed for some patients in the hospital. Two cancer types are targeted: prostate cancer and pancreas cancer. This choice is made based on data-availability and number of patients treated at the hospital. For these two types of cancer, patient records are analyzed based on the following criteria:

- All patients treated for the first time between January and June 2014 and coded in a Diagnosis-Related Group of prostate cancer or pancreas cancer
- Both inpatient and outpatient treatment were considered
- Patients who have been treated for cancer in the 3 previous years are excluded

Data comes from three sources: the medical records management system, the appointment scheduling system, and the Diagnosis-Related-Group payment management system. Following information is gathered:

- Age, ID, cancer code
- Date and department of the first appointment, date of announcement consult
- Cancer characteristics: metastatic or not, diagnosis (with biopsy results) available at first consult or not, additional exams needed or not, first Gleason score measured for prostate cancer (Gleason score measures the aggressiveness of a tumor. The higher the Gleason score, the worse the prognosis.)
- Date of the first multidisciplinary meeting where the case was discussed
- Date of the biopsy and date of the biopsy report, dates and types of treatments
- Date of apparition of metastasis if relevant, date of decease if relevant
- Number of consults, number of hospital stays, departments involved in the care

As this was the first study of this type, information is extracted manually.

4 Materials and Results

4.1 *Qualitative Study: Process Design and Implementation*

The analysis of the documents produced by the process design team and the interview with the oncology professor provide a good understanding of the care process that was designed and the way it was designed. The cancer care process itself is pictured in Fig. 1. It is structured in four phases: entrance in the system, diagnosis, treatment and "after-care". It specifies a set of guiding principles:

- Cancer announcement should be done in two steps: one for diagnosis announcement, and one for treatment announcement

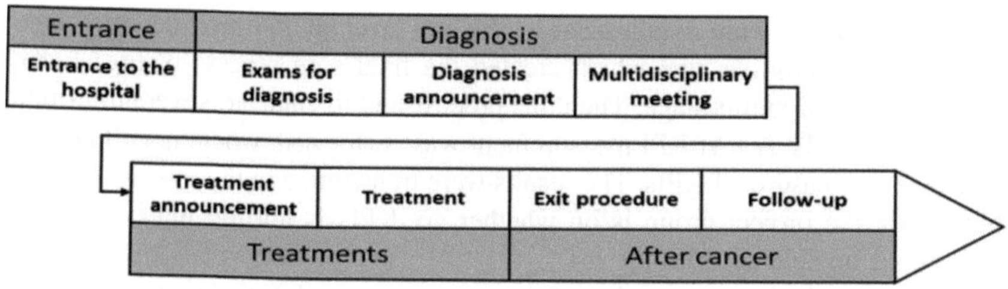

Fig. 1 Cancer care process

- These two announcement steps should be two different consults
- Between these two consults, treatment should be discussed in a multidisciplinary meeting, with at least three different medical specialties (e.g. urologist, oncologist and radiotherapist)

Six delay indicators are defined on this process. They are described in Table 2. This process is at a very high level. Figure 2 shows how it was designed.

First, all expert centers were asked to draw a model of the care process for their specialty (e.g. urology cancers or blood cancers). No common formalism was provided, so various level of details were obtained. All the processes modelled by

Table 2 KPIs for the cancer care process

#	From	To
D1	Scheduling of first appointment	First appointment
D2	First appointment	Multidisciplinary meeting (MDM)
D3	First appointment	First treatment
D4	Date of biopsy	Biopsy report
D5	Date of biopsy	Molecular biology report
D6	First treatment	Second treatment

Fig. 2 Process design project

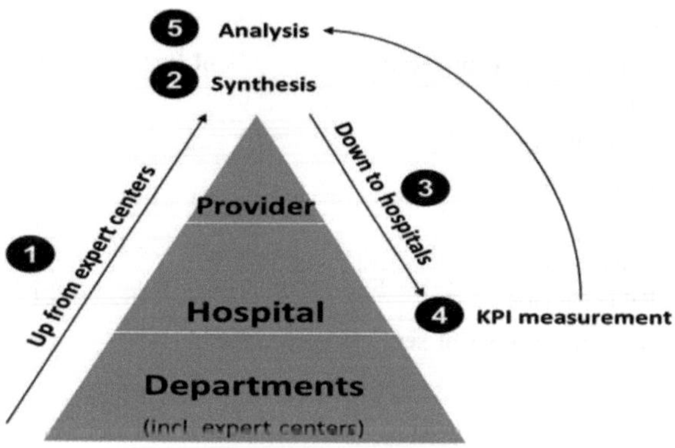

the expert centers were synthesized at the top-level of the provider in a mixed descriptive-prescriptive way which created the final care process of Fig. 1. KPIs were also defined at this stage. Then this process and its indicators were transmitted down to hospitals. A first KPI measurement was performed, where not all hospitals were able to measure all KPIs. The results were transmitted to the top level. Current thinking in the project group is on whether six KPIs is not too much and which KPIs could be deleted from the list.

Concerning the implementation of this process, in our hospital no resources have been allocated to this project. KPIs are not routinely computed. This will allow us to study the "real system" before implementation of this new procedure.

4.2 *Quantitative Study: Current Situation in the Hospital*

Two cancer types are studied: prostate and pancreas. For prostate cancer, 120 candidates are identified for inclusion. However, after the medical records have been analyzed, only 70 are included. For pancreas cancer, from 39 candidates 21 patients are included. Details are provided in Table 3.

These two populations are very different. They don't use the same resources: on average, pancreas cancer patients transit through 3.3 departments, which is more than twice the number of departments for prostate cancer patients.

The types of treatment are also very different: mostly surgery for prostate cancer patients, chemotherapy and palliative care for pancreas cancer patients. Survival rates are completely different.

Table 3 Type of treatment received for prostate cancer care

	Prostate		Pancreas	
Inclusion	**% of cand.**	#	**% of cand.**	#
Candidates for inclusion	100	120	100	39
Patients included	58	70	54	21
Wrong coding	23	27	28	11
Cancer in the previous 3 years	26	31	38	15
Type of treatment received	**% of incl.**	#	**% of incl.**	#
Surgery	91	64	10	2
Chemotherapy	3	2	52	11
Radiotherapy	4	3		0
Hormonotherapy	10	7		0
2 treatments	9	6	10	2
Palliative care	1	1	43	9
Average number of services	#		#	
	1.5		3.3	

Finally, ways of entrance in the hospital also differ: most prostate cancer patients are detected through a prevention scheme, whereas there is no such protocol for pancreas cancer and most patients come for abdominal pains or other symptoms that need to be related to pancreas cancer.

As the number of included patients is low for pancreas cancer, only prostate cancer will be analyzed further. Two interesting comparisons can be made: with the provider-level process, and with national data. Table 4 shows delays computed for prostate cancer patients in the hospital (hosp.) and corresponding delays from a national investigation undertaken by the national cancer institute, INCa [18]. This investigation included both general hospitals and university hospitals, from the public and private sector, in 17 French regions (which cover almost 50 % of the French population).

For mean and median computation, only positive values are included. They show that time from biopsy report to MDM is shorter in this hospital than nationwide. However, delays from MDM to surgery are longer in this hospital.

But the main comment that can be made on these results is on the ratio of negative results (the proportion of negative delays for all included patients). For the delay from MDM to surgery, this ratio is of 55 % in this hospital, and 25 % nationwide. It means that most patient get their surgery, and their case is discussed in the MDM only after that, when treatment has already been performed. This goes against the recommendations of the provider-level process. It also appears that surgery rates are much higher in this hospital (91 % of patients) than nationwide (49 % of cases in the INCa study).

Some additional treatment can be performed. For patients who had surgery, if no biopsy is performed at the hospital (which means that they already had biopsy results when they arrived), the delay between the first appointment and the treatment is significantly shorter (Student's t-test, unilateral, $p < 0.001$, 38 observations with biopsy at the hospital, 32 without). Also the relation between Gleason score

Table 4 Delays for prostate cancer care (MDM: multidisciplinary meeting)

		For all patients	If surgery is performed	
		Biopsy report to MDM (days)	MDM to surgery (days)	Biopsy report to surgery (days)
Number	Hosp.	70	63	63
	INCa	3050	1353	1350
Data availability (%)	Hosp.	96	44	45
	INCa	73	66	66
Mean (SD)	Hosp.	10 (49)	109 (92)	122 (102)
	INCa	37 (34)	45 (30)	81 (37)
Median (interquartile)	Hosp.	7 (5–9)	77 (38–143)	89 (50–155)
	INCa	29 (14–53)	39 (23–61)	77 (56–103)
Ratio of negative results (%)	Hosp.	0	55	55
	INCa	3	25	9

and time to treatment can be analyzed. It shows that for Gleason scores 8, 9 and 10 (the most aggressive tumors), the delay between first appointment and treatment is significantly shorter than for Gleason scores 7 (Students t-test, unilateral, $p < 0.01$, 51 observations with Gleason 7, 12 with Gleason above 7). Therefore, if the diagnosis is clear from the beginning, time to treatment is faster, and the more serious the tumor the shorter the time to treatment. This is what would be expected.

5 Discussion

5.1 Global Project and Local Specificities

Although patient record analysis is a long process (about a full week of work for one person to analyze the 159 records), a lot of data can be extracted. However, it is not clear how to make sense of this data.

A first point is surgery rates for prostate cancer. It is much higher in this hospital than nationwide. Nevertheless, the national study includes patients who were in the first steps of cancer and were put under surveillance. These “early” patients don’t get hospitalized.

A second point is the rate of negative delays between MDM and treatment. One reaction could be to blame the hospital studied here: it is not following recommendations. However, this hospital has a well-known expertise for prostate surgery. It has some very specific equipment that make it a reference center for this type of intervention. Probably people come to this hospital for this reason: they choose surgery as a treatment with their private practitioner, and then go to this famous reference center. The surgeon only follows the patient’s decision, for a surgery that has become almost a routine operation. These cases may also have been discussed in a MDM in another organization, after which the patient chose to be treated in this reference center.

Delays are another element to look at. Time to surgery can be longer or shorter (here it is difficult to say due to the high rate of negative delays), but it will depend on the population addressed by the hospital. The performed analysis shows that the more aggressive the tumor, the shorter the time to treatment. Delays depend on the severity of the patient’s condition, which is understandable. Hospitals should therefore not be compared on this indicator. But even for one hospital, is this indicator supposed to rise or decrease? Actually, setting an objective could lead to counter-productive measures where the time from MDM to surgery is reduced on average but increased for more urgent patients.

All these elements (negative delays from MDM to surgery and high surgery rates due to hospital specialization and reputation, delays whose interpretation is uncertain) point to the specificity of each situation. Local contexts need to be taken into account to create relevant processes and indicators. Characteristics of this local context that need to be paid attention to include:

- Hospital specialization in one technique or another
- Hospital or physician reputation
- Population addressed
- Type of organization

Here, indicators could be associated to objectives, but these objectives would not reflect the complexity of the situation.

The comparison of prostate cancer and pancreas cancer is also instructive. It shows that treatments are very different for these two types of cancer, in the same organization: surgery is the main treatment for prostate cancer in the studied hospital, chemotherapy or palliative care for pancreas cancer.

The analysis performed and a discussion with practitioners showed that the uncertainty of the diagnosis is very different in these two cases. Pancreas cancer can go unnoticed for a while, and patients often come to the hospital with an advanced cancer. On the other hand, prostate cancer evolves more slowly, it has a nationwide detection scheme, and prostate surgery seems to have become a more routine operation in this hospital due to the large number of cases treated.

Therefore disease complexity should also be considered when establishing processes or indicators. Procedures can be more or less complex: standard, routine or non-routine [19]. Van der Geer et al. [20] have shown that when they are given the opportunity to develop their own performance indicators, medical teams obtain different results depending on the uncertainty of the tasks and diseases. Therefore defining common indicators for these two types of cancer appears very challenging as the activities measured should not be the same: problem-solving for pancreas-cancer, treatment and outcomes for prostate cancer.

Vertical (across hierarchies), horizontal (across domain specialties) and longitudinal (along time) integration are needed [8] and should always take into consideration the complexity of local operations. Complexity can be described in three dimensions: diversity, multiplicity and interconnectedness [21]. In this case multiplicity has been considered but the importance of the diversity of patients, disease characteristics and organizations has been underestimated. Besides, interconnectedness between organizations characteristics and patient profiles has also been forgotten.

These considerations are at the heart of complex systems engineering. One of the conclusions is that systems engineering could help as it starts with the analysis of goals and context, then specifies the system before actually designing it. Here context analysis and goals of different shareholder have not been given enough attention which is why results are not as useful as hoped. When gathering information on current practices, specifications on process models were not given to expert centers, which is one of the reasons why the final unified process model was hard to design.

Figure 3 shows a proposition of integration scheme for cancer care in the provider studied here. It differs from the current architecture:

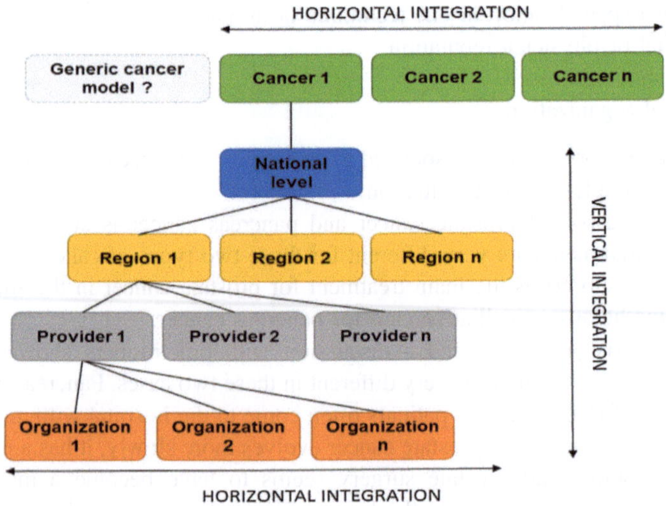

Fig. 3 Vertical and horizontal integration for cancer care in France: proposed framework

- Cancer types are differentiated at the top level, as they constitute different diseases with different care strategies and challenges. Horizontal integration must be achieved types if a generic metric for cancer care is desired.
- Comparisons between lower-level elements are possible at each level, but the metrics are not necessarily the same at each level.

This is related to the difference between tall organizational complexity and symmetric organizational complexity defined by Burton et al. [22]. Tall organizational complexities are fit for functional configurations in varied but predictable environments. Symmetric complexity is needed for turbulent (varied and hardly predictable) environments with matrix organizations: here, medical departments and care process across these departments.

5.2 *Strengths and Limitations*

In this study, medical records for a 6-months period are analyzed. This analysis provides insights on how operations are currently performed, and how they are thought of at the top-level: the "bottom-up view" of a "top-down" initiative.

However, the low number of patients for pancreas cancer did not allow much analysis on this population. Almost half of the candidate pancreas cancer patients could not be included. This relates to the more global issue of data access. This problem has already been identified in France [18] and other countries [23]. Data is hard to access as it is available in the form of free-text in medical records rather than in a database. It is also scattered between different information systems, and the reliability of data in these systems is not optimal as wrong coding is common.

5.3 *Future Developments*

This study underlines the shortcomings of centralized, top-down approaches to the management and coordination of a complex system such as a group of hospitals. In this multi-level complex system, local situations are different and require differentiated approaches. Complex organizational systems theory and network analysis can be of great help.

The analysis of patient journeys for prostate cancer care and pancreas cancer care has shown that inside each cancer type different trajectories coexist. However, these trajectories are not readily available: patient data needs to be analyzed so that relevant patient groups can be identified. Data analysis methods such as Classification And Regression Tree analysis have already been used for operations research in healthcare [24]. They could be used to cluster patients for care coordination. For each of these groups of patients, an assessment of the coordination needs of the patients and the organization needs to be performed, so that appropriate resources can be allocated and other actions can be taken (reshape multidisciplinary teams, work on information systems…).

Finally, for such large systems as this health provider, "handmade" process mining is not enough and automated methods must be developed. However, hospital information systems are particularly divided and make this quite difficult.

References

1. OECD: Health at a glance 2013: OECD indicators. Organisation for Economic Co-operation and Development, Paris (2013)
2. Cour des Comptes: La Sécurité sociale—Rapport sur l'application des lois de financement de la sécurité sociale. Cour des Comptes, Paris, France (2014)
3. Plsek, P.E., Greenhalgh, T.: The challenge of complexity in health care. BMJ **323**, 625–628 (2001)
4. Georgopoulos, B.S., Matejko, A.: The American general hospital as a complex social system. Health Serv. Res. **2**, 76–112 (1967)
5. Glouberman, S., Mintzberg, H.: Managing the care of health and the cure of disease–part II: integration. Health Care Manage. Rev. **26**, 70–84; discussion 87–89 (2001)
6. Reid, P.P., Grossman, J.H. (eds.): A framework for a systems approach to health care delivery. In: Building a better delivery system: A new engineering/health care partnership. National Academies Press, Washington, DC, États-Unis (2005)
7. Dupuy, F.: Lost in management: la vie quotidienne des entreprises au XXIe siècle. Éditions du Seuil, Paris (2011)

8. Nyssen, A.-S.: Coordination in hospitals: organized or emergent process? Cogn. Technol. Work **9**, 149–154 (2007)
9. McDonald, K.M., Sundaram, V., Bravata, D.M., Lewis, R., Lin, N., Kraft, S.A., McKinnon, M., Paguntalan, H., Owens, D.K.: Closing the quality gap: a critical analysis of quality improvement strategies, vol. 7: Care Coordination. Agency for Healthcare Research and Quality (US), Rockville (MD) (2007)
10. Vera, A., Kuntz, L.: Process-based organization design and hospital efficiency. Health Care Manage. Rev. **32**, 55–65 (2007)
11. Axelsson, R., Axelsson, S.B., Gustafsson, J., Seemann, J.: Organizing integrated care in a university hospital: application of a conceptual framework. Int. J. Integr. Care. **14**, e019 (2014)
12. Ben-Tovim, D.I., Dougherty, M.L., O'Connell, T.J., McGrath, K.M.: Patient journeys: the process of clinical redesign. Med. J. Aust. **188**, S14–S17 (2008)
13. Vanhaecht, K.: The Impact of Clinical Pathways on the Organisation of Care Processes (2007)
14. Panella, M., Marchisio, S., Di Stanislao, F.: Reducing clinical variations with clinical pathways: do pathways work? Int. J. Qual. Health Care **15**, 509–521 (2003)
15. Dy, S.M., Garg, P., Nyberg, D., Dawson, P.B., Pronovost, P.J., Morlock, L., Rubin, H., Wu, A.W.: Critical pathway effectiveness: assessing the impact of patient, hospital care, and pathway characteristics using qualitative comparative analysis. Health Serv. Res. **40**, 499–516 (2005)
16. van de Klundert, J., Gorissen, P., Zeemering, S.: Measuring clinical pathway adherence. J. Biomed. Inform. **43**, 861–872 (2010)
17. Walden, D.D., Roedler, G.J., Forsberg, K., Hamelin, R.D., Shortell, T.M.: International council on systems engineering eds: systems engineering handbook: a guide for system life cycle processes and activities. Wiley, Hoboken (2015)
18. INCa: Délais de prise en charge des quatre cancers les plus fréquents dans plusieurs régions de France en 2011 et 2012 : sein, poumon, côlon et prostate. Institut National du Cancer, Paris, France (2013)
19. Lillrank, P., Liukko, M.: Standard, routine and non-routine processes in health care. Int. J. Health Care Qual. Assur. Inc. Leadersh. Health Serv. **17**, 39–46 (2004)
20. Van der Geer, E., van Tuijl, H.F.J.M., Rutte, C.G.: Performance management in healthcare: performance indicator development, task uncertainty, and types of performance indicators. Soc. Sci. Med. **69**, 1523–1530 (2009)
21. Jacobs, M.A.: Complexity: toward an empirical measure. Technovation **33**, 111–118 (2013)
22. Burton, R.M., DeSanctis, G., Obel, B.: Organizational design a step-by-step approach. Cambridge University Press, Cambridge (2006)
23. Zegers, M., de Bruijne, M.C., Spreeuwenberg, P., Wagner, C., Groenewegen, P.P., van der Wal, G.: Quality of patient record keeping: an indicator of the quality of care? BMJ Qual. Saf. **20**, 314–318 (2011)
24. Harper, P.R.: A framework for operational modelling of hospital resources. Health Care Manag. Sci. **5**, 165–173 (2002)

Computational Intelligence Based Complex Adaptive System-of-System Architecture Evolution Strategy

Siddhartha Agarwal, Cihan H. Dagli and Louis E. Pape II

Abstract There is a constant challenge to incorporate new systems and upgrade existing systems under threats, constrained budget and uncertainty into systems of systems (SoS). It is necessary for program managers to be able to assess the impacts of future technology and stakeholder changes. This research helps analyze sequential decisions in an evolving SoS architecture through three key features: SoS architecture generation, assessment and implementation through negotiation. Architectures are generated using evolutionary algorithms and assessed using type II fuzzy nets. The approach accommodates diverse stakeholder views, converting them to key performance parameters (KPPs) for architecture assessment. It is not possible to implement an acknowledged SoS architecture without persuading the systems to participate. A negotiation model is proposed to help the SoS manger adapt his strategy based on system owners' behavior. Viewpoints of multiple stakeholders are aggregated to assess the overall mission effectiveness of an architecture against the overarching objectives. A search and rescue (SAR) example illustrates application of the method. Future research might include group decision making for evaluating architectures.

Keywords Architecture · Acquisition · Evolutionary algorithms · Machine learning · Systems of systems · Meta-Architectures

S. Agarwal (✉) · C.H. Dagli · L.E. Pape II
Missouri University of Science and Technology, Rolla, MO 65409, USA
e-mail: sa265@mst.edu

C.H. Dagli
e-mail: dagli@mst.edu

L.E. Pape II
e-mail: lep7df@mst.edu

G. Auvray et al. (eds.), *Complex Systems Design & Management*,
DOI 10.1007/978-3-319-26109-6_9

1 Introduction

In the real world, systems are complex, non-deterministic, evolving, and have human centric behaviors. The connections between complex systems are non-linear, globally distributed, and evolve both in space and in time. Because of non-linear properties, system connections create an emergent behavior. It is imperative to develop an approach to deal with such complex large-scale systems. Complex entities include both socioeconomic and physical systems, which undergo dynamic, rapid changes. Some examples of complex systems include transportation systems [1], health systems [2], internet of things [3], defense frameworks [4], and manufacturing infrastructures [5], among others.

Another recently emerged concept is Cyber Physical Systems (CPS) [6]. A CPS is a SoS which integrates physical system with cyber capability to improve the performance [7]. Cyber capability includes models of the process that can be used to make decisions over the system.

Classical system architecting deals with static systems, but the process of System of Systems (SoS) architecting should be started at a meta-level. The meta-architecture sets the tone of the architectural focus [8] and drives the process of architecting further. A meta-architecture provides multiple alternatives for the final architecture. SoS architecting integrates multiple systems' architectures to produce an overall, large scale system meta-architecture for a specifically designated mission [9].

Architecture simulation and modeling techniques for Acknowledged SoS are still in their early stages. Recent works in this area include: DANSE, standing for Designing for Adaptation and Evolution in System of Systems [10]. That project addresses the challenging technical, management, and political problems within organizations. The DYMASOS (Dynamic Management of Physically Coupled Systems of Systems) Project explores methods for the distributed management of large, physically connected systems along with distributed autonomous management and global coordination [11]. COMPASS is the Comprehensive Modeling for Advanced Systems of Systems. It aims to develop collaborative research on model-based techniques for developing and maintaining SoS [12].

This research is the first attempt to combine multiple behaviors of systems participating in a complex adaptive SoS operational scenario. It proposes the use of neural networks [13] to help the SoS manager adapt his negotiation strategy while dealing with multiple constituent systems on multiple issues such as deadline, funding or performance. Our attempt here is to present an integrated acknowledged SoS architecting model whose capabilities include SoS meta-architecture generation covering the entire design space, flexible and robust architecture assessment, and final architecture implementation through simulated negotiations.

Flexible and Intelligent Learning Architectures for SoS (FILA-SoS) is the name of this process. The FILA-SoS has several independent modules that may be used together for meta-architecture generation, architecture assessment, meta-architecture executable modeling, and architecture implementation through negotiation.

Architecture generation methods include fuzzy-genetic [14], multi-level [15], particle swarm [16] and cuckoo search optimization [16]. The initial architecture assessment method was based on type-1 fuzzy logic systems (FLS) [17].

Implementing an acknowledged SoS architecture requires persuading the systems to participate. To address this issue, a negotiation protocol is based on game theory [18]. Individual systems may use three kinds of negotiation models based on their negotiation strategies: a non-cooperative Linear Optimization model, a cooperative fuzzy negotiation model, and a semi-cooperative Markov chain model. Executable architectures are generated using a hybrid of Object Process Methodology (OPM) and Colored Petri Nets (CPN) [16, 19, 20]. The evolution of the SoS should take into account availability of legacy systems and new systems willing to join, adaptation to changes in mission, and sustainability of the overall operation [21, 22].

In this paper three new modules are introduced for FILA-SoS: an alternative meta-architecture generation model based on swarm intelligence, a new architecture assessment technique based on type-II fuzzy logic, and a bilateral negotiation mechanism for one SoS manager with many individual systems based on clustering and machine learning techniques. These modules can help in designing an overall evolution strategy for complex adaptive SoS (CASoS). The proposed approach is implemented through a notional Coast Guard Search and Rescue (SAR) problem serving the Alaskan Coast Region shown in Fig. 1.

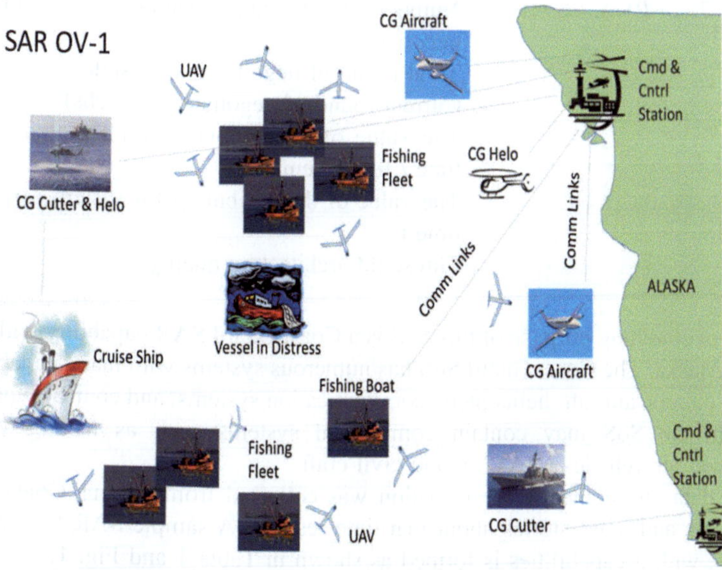

Fig. 1 Operational view (OV-1) for search and rescue scenario

2 FILA-SoS Integrated Model Variables and Parameters

$\boldsymbol{C}$	The overall capability (the overall SoS goal to be achieved by combining sub-capabilities from systems)
$c_j : j \in \boldsymbol{J}, \boldsymbol{J} = \{1, 2, \ldots, M\}$	Constituent system capabilities j required to achieve $\boldsymbol{C}$
$S_i : i \in \boldsymbol{I}, \boldsymbol{I} = \{1, 2, \ldots, N\}$	Candidate system i for the SoS
N	Total number of systems candidates
M	Total number of capabilities required

Let $\boldsymbol{A}$ be a $N\text{x}M - matrixofa_{ij}where$
$a_{ij} = 1$ *if* capability j is possessed by system i
$a_{ij} = 0$ *otherwise*
P_i Performance of system i for delivering all its capabilities
F_i Funding of system i for delivering all its capabilities
D_i Deadline to participate in this round of mission development for system i

IF_{ik} *is the* interface between systems i *and* k *s.t.* $i \neq k, k \in I$

IC_i	The cost for development of interface for system i
OC_i	The cost of operations for system i
$KP_r : r \in \boldsymbol{R}, R = \{1, 2, \ldots, Z\}$	The key performance attribute r of the SoS
FA	Funding allocated to SoS Manager
$p = \{1, 2, \ldots, P\}$	Number of negotiation parameters for bilateral negotiation
t_{max}	Total round of negotiations possible
t	Current round of negotiation (epochs)
$V_{pi}^{SoS}(t)$	The value of the attribute p for SoS manager at time t for system i
$V_{pi}^{S}(t)$	The value of the attribute p for system i owner at time t
TQ	Threshold architecture quality

The overarching purpose of this SoS is a Coast Guard SAR capability within the Sea of Alaska. The Coast Guard SoS has numerous systems with many capabilities, such as cutters, aircraft, helicopters, communication systems, and control centers. In addition, the SoS may contain commercial systems such as fishing vessels, unmanned air vehicles (UAV's), and civil craft.

The data and contextual information was collected from various Coast Guard documents and news stories about maritime rescues. A sample SAR SoS with 22 systems, with 5 capabilities is formed as shown in Table 1 and Fig. 1.

Table 1 Types of the systems and capabilities present in the SoS

Sys. No.	Type	Cap.	Cap. No.	Capability name
s_1 and s_2	Cutter	2,5	c_2	High speed
s_3 and s_4	Helicopter	2,5	c_2	High speed
s_5 and s_6	Aircraft	2,5	c_2	High speed
s_7 to s_{12}	UAV	1,5	c_1	IR and night vision
s_{13} to s_{16}	Ship or Vessel	3,5	c_3	Deliver medical aid
s_{17} and s_{18}	Coordination/Control	4,5	c_4	RF direction finding
s_{19} to s_{22}	Communication	5	c_5	Communication systems

3 Meta-Architecture Generation

In a SoS architecting problem, component systems have multiple intra and inter system trade-offs that cannot be fit into the mold of a single objective. Secondly, the number of solutions required for approximation increases exponentially with the dimensionality of the objective space [23]. The SoS architect's aim is to optimize objective functions KP_r, as the case may be.

The SoS optimization problem can be formulated as follows:

Optimize $F = \{f_{KP_1}(s, IF), \ldots, f_{KP_r}(s, IF), \ldots, f_{KP_Z}(s, IF)\} \quad \forall r = \{1, 2, \ldots, Z\}$ Where $f_{KP_r}(s, IF)$ is the value of the key performance attribute r for decision variables s and IF Subject to

$$\sum_i s_i a_{ij} \geq 1 \quad \forall j \in J \tag{1}$$

$$IF_{ik} = \{1\} \leftrightarrow \{s_i = 1 \wedge s_k = 1\} \quad \forall i, k \in I \tag{2}$$

$$a_{ij} \in \{0, 1\} \quad \forall i \in I \tag{3}$$

$$s_i \in \{0, 1\} \quad \forall i \in I \tag{4}$$

$$IF_{ik} \in \{0, 1\} \quad \forall i, k \in I \tag{5}$$

This is a Z dimensional multi-objective optimization problem. Constraint (1) guarantees that at least one system for each capability is selected. Constraint (2) insures that an interface between two systems is selected if and only if the two systems are selected in the meta-architecture. Constraints (3, 4) give the binary decision variables. A similar problem was solved earlier as a multi-level bi-objective optimization [15] using gradient based methods. The bi-objective model cannot handle as many objectives as the general model described here.

There are three basic issues to be addressed: ambiguity in the definition of the key performance attributes (KPA), the number of objectives, and NP completeness of the mathematical model formulated. In this research, evolutionary algorithms (EA) using non-gradient descent optimization are selected to deal with the NP completeness issues; fuzzy logic is used to represent the ambiguity in KPA; and fuzzy inference is used to accommodate many objectives in formulating the fitness function. Fuzzy logic also helps in the search ability of the EA, since search ability decreases with increasing objectives [24]. The model is converted to a form where any EA can be used. Each chromosome is coded as a finite length vector of variables. The possible values of the variables equal the size of the alphabet. In this case the size of the alphabet is two because s_i *and* IF_{ik} are the binary decision variables.

The information for variables such as performance of each system P_i, funding allocated to each system F_i, deadline for preparation D_i, interface cost IC_i, and operations cost OC_i can be found from Table 2. There are five key performance attributes selected for this SoS:

$KP_1 = P^{SoS}$: Performance of SoS
$KP_2 = A^{SoS}$: Affordability of SoS
$KP_3 = R^{SoS}$: Robustness of SoS
$KP_4 = M^{SoS}$: Modularity of SoS
$KP_5 = NC^{SoS}$: Net-Centricity of SoS

Table 2 Domain data inputs for modeling the meta-architecture

Sys. No.	Type	Capability	*ICi*	*OCi*	*Pi*	*Di*	*LYi*	*SPi*
1	Cutter	2	0.03	0.2	12	1	8.3	6
2	Cutter	2	0.03	0.2	12	1	8.3	6
3	Helicopter	2	0.1	0.2	20	1	10.0	8
4	Helicopter	2	0.1	0.2	20	1	10.0	8
5	Aircraft	2	0.1	0.5	10	1	10.0	10
6	Aircraft	2	0.1	0.5	10	1	10.0	10
7	UAV	1	0.1	0.1	7	1	1.7	2
8	UAV	1	0.1	0.1	7	1	1.7	2
9	UAV	1	0.1	0.1	7	1	1.7	2
10	UAV	1	0.1	0.1	7	1	1.7	2
11	UAV	1	0.1	0.1	7	1	1.7	2
12	UAV	1	0.1	0.1	7	1	1.7	2
13	Fish Vessel	3	0.03	0.5	10	1	5.0	4
14	Fish Vessel	3	0.03	0.5	10	1	5.0	4
15	Fish Vessel	3	0.03	0.5	10	1	5.0	4
16	Civ Ship	3	0.05	2	8	1	6.7	4
17	Coord. Ctr.	4	0.05	0.5	5	1	0.5	0
18	Coord. Ctr.	4	0.05	0.5	5	1	0.5	0
19	Comm.	5	0.02	0.03	1	0	0.5	0
20	Comm.	5	0.02	0.03	1	0	0.5	0
21	Comm.	5	0.02	0.03	1	0	0.5	0
22	Comm.	5	0.02	0.03	1	0	0.5	0

Related information required for SoS architecture generation is LV_i: the systems performance among participating systems based on ability to search and provide assistance and SP_i: the systems' speeds (Table 2). Three negotiation attributes for bilateral negotiation are Funding, Deadline, and Performance.

Modular fuzzy net process is used for to assessing the fitness of the of individual architecture instances (chromosomes). First, we calculate the values of inputs which depend on the architecture instance required for each KPA (e.g., affordability, performance, net-centricity, etc.). Crisp values for the KPAs are calculated using Type I fuzzy rules based on the stakeholder's views. For example, a rule can be written stating "If operations cost is high and the interfacing cost is high, then affordability is low." These fuzzy rules are used to assign a crisp number to the affordability of the overall architecture. Each of the KPAs are then modeled as interval type II fuzzy sets (IT2FS) so that a crisp value can be obtained for the architectures' overall quality.

A modular fuzzy net process is used for assessing the fitness of the of individual architecture instances (chromosomes). First, we calculate the values of inputs that are required for each KPA (e.g., affordability, performance, and net-centricity). Crisp values for the KPAs are then calculated using Type I fuzzy rules. These rules are based on the stakeholder's views. For example, a rule can be written that states the following: "If operations cost is high and the interfacing cost is high, then affordability is low". Each of the KPAs are then modeled as interval type II fuzzy sets (IT2FS) so that a crisp value can be obtained for the architectures overall quality.

IT2FSs have been shown to be more capable of modeling uncertainties than are Type 1 FSs. Each KPA with its inputs is referred to as a module. Type I FSs are used within modules to reduce computational time. The rules of the fuzzy evaluator are adjustable to allow for differences between the stakeholders' views. This adjustability makes fuzzy nets usable for a larger set of similar domain problems, as well as being applied to other domains. The fuzzy network helps control uncertainties at lower levels of the KPA. KPAs of the SoS can be provided with different levels of linguistic granularization such as:

Affordability: very costly, costly, cheap
Modularity: little, average, good
Performance: very low, mediocre, great
Robustness: less, ordinary, excellent
Net Centricity: low, medium, high

Triangular type-2 membership functions were used for all attributes. Twenty-five rules were created to link these five objectives to four fuzzy attributes. These statements help clarify stakeholders' perspectives (Fig. 2).

Particle swarm optimization, inspired by the behavior of bird flocks or fish schools [25], is used with a fuzzy evaluator. Figure 3 shows the results.

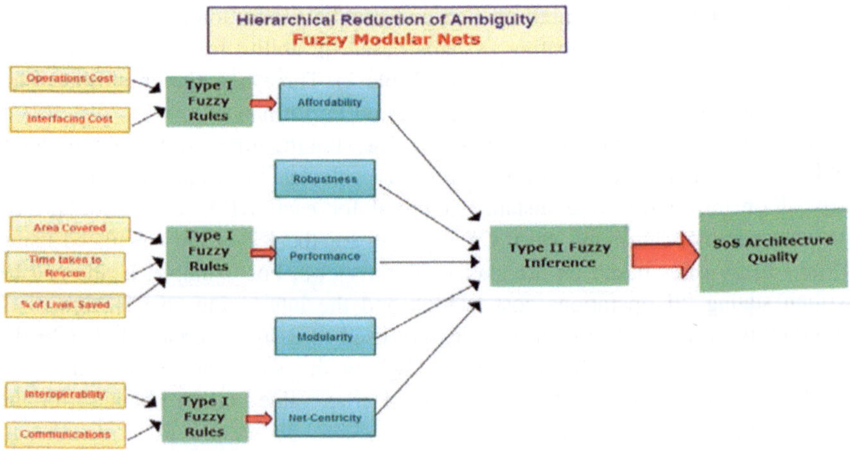

Fig. 2 The fuzzy nets to evaluate architecture's quality

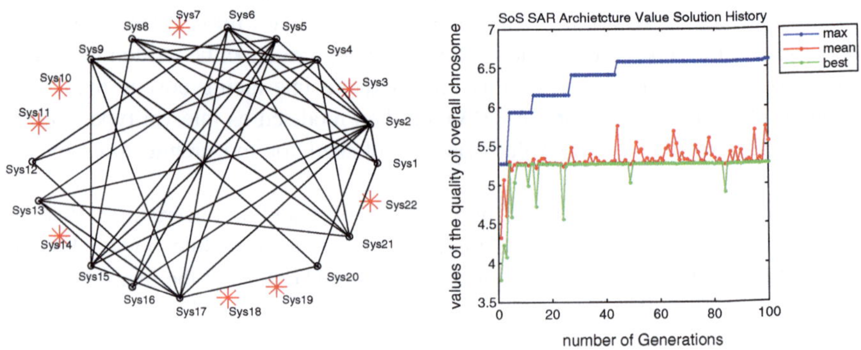

Fig. 3 Systems selected in the SAR-22 SoS architecture through BPSO

4 SoS Meta-Architecture Negotiation

The Acknowledged SoS manager negotiates with systems that are selected as part of a desirable architecture during the meta-architecture generation process. A negotiation procedure is necessary for the realization or implementation of the meta-architecture generated. Since a SoS manager cannot force his demands on participating systems, negotiation helps in achieving an architecture that is implementable. The SoS manager negotiation mechanism consist of three phases:

(i) Modelling the opponent
(ii) Making a decision based on the previous offer
(iii) Generating a counteroffer.

A bilateral counteroffer based negotiation mechanism under multiple attributes as depicted in Fig. 4 is chosen. The attributes are assumed to be independent and are

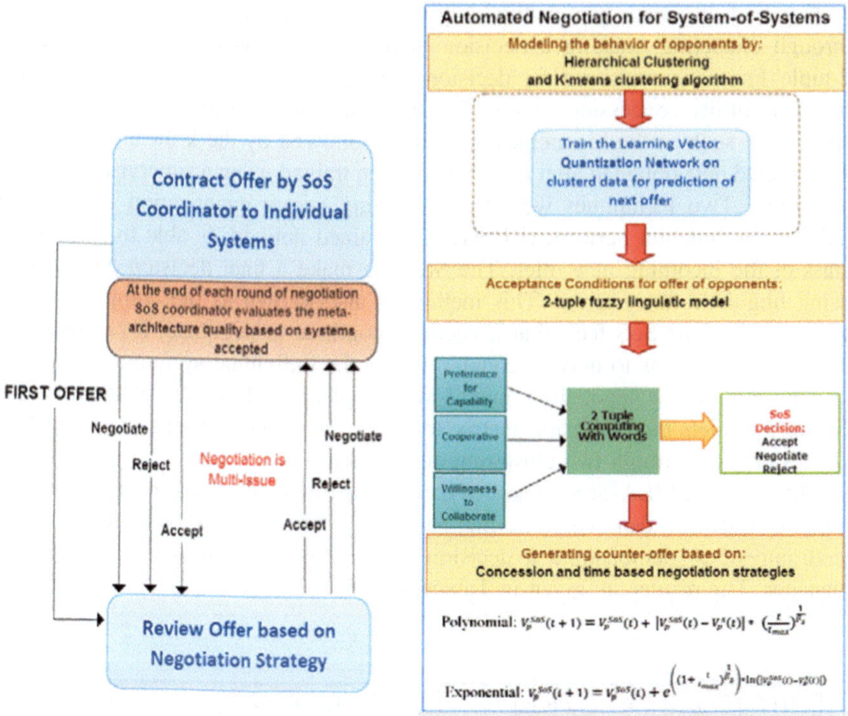

Fig. 4 Three salient features of automated negotiation

bargained simultaneously. Modeling the opponent characterizes their negotiation behaviors, selected from among: cooperative, semi-cooperative or non-cooperative. A decision mechanism rejects the offer for no further negotiation, or accepts the offer as it is currently, or negotiates for another round. In case of further negotiations, a counter-offer generation mechanism is used. Counter-offers in automated negotiation are classified on the basis of bargaining constraints, such as remaining time to negotiate, value of the overall utility achieved by a party, or constraints based on available resources. Figure 4 gives an overview of the three salient features of automated negotiation used [26]. The strategy is designed for a one-to-many negotiation problem. It is not mediated by a coordinator. The structure consists of a SoS manager and multiple selected systems from the meta-architecture. Defining: V_p: $p = \{1, 2,\ldots, P\}$: Then attributes for bilateral negotiation are

t_{max}: Total round of negotiations possible $t = \{0, 1, \ldots, t_{max}\}$
$V_p^{SoS}(t)$: The value of the attribute V_p for SoS manager at time t
$V_p^S(t)$: The value of the attribute V_p for system owner at time t.

A number of negotiation rounds with different system types and SoS coordinator are conducted. Negotiation offers made by systems reveal incomplete information about their preference of issues and their strategy.

Figure 4 describes the methodology for modeling the opponents' behavior through clustering, making a decision on the negotiation offer based on fuzzy 2-tuple linguistic multi-criteria decision making, and generating a counteroffer based on utility concession curves. The figure explains the processes involved in succession such as the hierarchical clustering followed by the k-means clustering. The labeled data obtained after clustering is then trained using a supervised learning algorithm. Two techniques were tried: learning vector quantization (LVQ) and radial basis function network (RBFN). The trained network is able to predict the class of the incoming new offer. The SoS can make a final decision on the offer using linguistic fuzzy terms. This method is also known as the computing with words [27]. If the SoS feels that it needs to continue to negotiate, it can use time dependent equations to make a counteroffer to the individual systems.

This automated negotiation model was implemented on the SAR problem. Figure 5 depicts the confusion matrix for neural networks training based on four classes created through two clustering algorithms.

The modified FILA-SoS integrated model was run for three waves to understand how the SAR SoS can evolve in time based on different scenarios imposed by the environment to demonstrate a decision making tool for an Acknowledge SoS manager. The results are given in Tables 3, 4, 5, 6, 7 and 8.

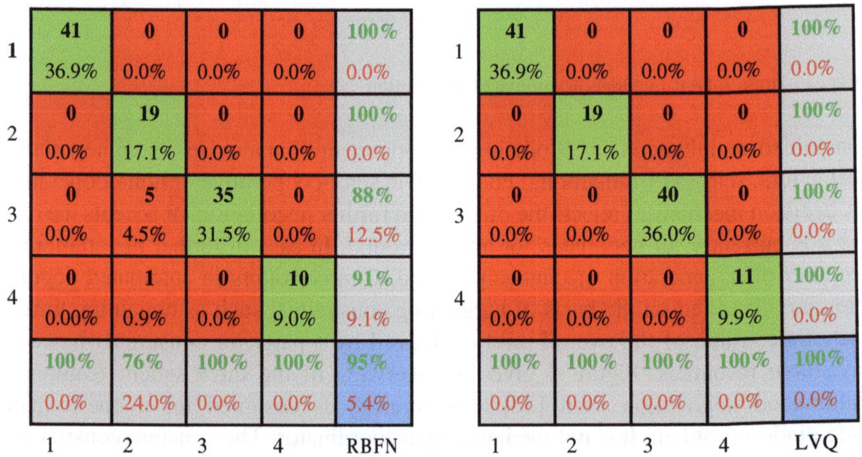

Fig. 5 Confusion matrices for training and testing by both RBFN and LVQ

Table 3 Meta-Architecture wave 1

Quality	3.11
Performance	3.36
Affordability	3.01
Net-Centricity	2.55
Robustness	2.74

Table 4 Negotiated-Architecture wave 1

Quality	1.75
Performance	2.8
Affordability	3.7
Net-Centricity	1.55
Robustness	1.74

Table 5 Meta-Architecture wave 2

Quality	3.29
Performance	3.21
Affordability	2.98
Net-Centricity	3.64
Robustness	3.74

Table 6 Negotiated-Architecture wave 2

Quality	2.12
Performance	1.8
Affordability	2.58
Net-Centricity	2.07
Robustness	1.33

Table 7 Meta-Architecture wave 3

Quality	3.21
Performance	3.09
Affordability	3.78
Net-Centricity	3
Robustness	2.79

Table 8 Negotiated-Arch wave 3

Quality	1.82
Performance	2.8
Affordability	3.7
Net-Centricity	1.55
Robustness	1.74

5 Concluding Remarks

The goal of this research is to model the evolution of the architecture of an acknowledged SoS, accounting for the ability and willingness of constituent systems to support the SoS capability development. The Wave Process Model provides a framework for modeling methodology, and this research provides different sets of modules to be integrated with the existing FILA-SoS. The research achieved the objectives to develop a simulation for acknowledged SoS architecture selection and evolution, with a structured, repeatable approach for planning, modeling, studying and evaluating the impact of individual system behavior on SoS capability and architecture evolution process. Results have been satisfactory and proved the model as a prototype.

Acknowledgment This material is based upon work supported, in whole or in part, by the U.S. Department of Defense through the Systems Engineering Research Center (SERC) under Contract H98230-08-D-0171. SERC is a federally funded University Affiliated Research Center managed by Stevens Institute of Technology. Any opinions, findings, conclusions, or recommendations expressed in this material are those of the author(s) and do not necessarily reflect the views of the United States Department of Defense.

References

1. Trentesaux, D., Knothe, T., Branger, G., Fischer, K.: Planning and control of maintenance, repair and overhaul operations of a fleet of complex transportation systems: a cyber-physical system approach. In: Service Orientation in Holonic and Multi-agent Manufacturing, pp. 175–186. Springer International Publishing (2015)
2. Obal, L., Lin, F.: A framework for healthcare information systems: exploring a large system of systems using system dynamics. Commun. IIMA **5**(3), 4 (2015)
3. Maia, P., Cavalcante, E., Gomes, P., Batista, T., Delicato, F. C., Pires, P. F.: On the development of systems-of-systems based on the internet of things: a systematic mapping. In: Proceedings of the 2014 European Conference on Software Architecture Workshops, p. 23. ACM, August 2014
4. Martí, J., Ventura, C., Hollman, J., Srivastava, K., Juarez, H.: I2Sim modelling and simulation framework for scenario development, training, and real-time decision support of multiple interdependent critical infrastructures during large emergencies. In: NATO (OTAN) MSG-060 Symposium on "How is Modelling and Simulation Meeting the Defence Challenges out to 2015?" (2015)
5. Nahavandi, S., Creighton, D., Le, V.T., Johnstone, M., Zhang, J.: Future integrated factories: a system of systems engineering perspective. In: Integrated Systems: Innovations and Applications, pp. 147–161. Springer International Publishing (2015)
6. Zhang, L.: Applying system of systems engineering approach to build complex cyber physical systems. In: Progress in Systems Engineering, pp. 621–628. Springer International Publishing (2015)
7. Dong, P., Han, Y., Guo, X., Xie, F.: A systematic review of studies on cyber physical system security. Int. J. Secur. Appl. **9**(1), 155–164 (2015)

8. Malan, R., Bredemeyer, D.: Architecture resources, defining non-functional requirements (2001)
9. Jamishidi, M.: System of systems-innovations for 21st century. In: IEEE Region 10 and the Third international Conference on Industrial and Information Systems, 2008, ICIIS 2008, pp. 6–7. IEEE, December 2008
10. Arnold, A., Boyer, B., Legay, A.: Contracts and behavioral patterns for SoS: the EU IP DANSE approach. arXiv preprint arXiv:1311.3631 (2013)
11. Paulen, R., Engell, S.: DYMASOS—dynamic management of physically coupled systems of systems, published on ERCIM News 97, April 2014, Special theme: Cyber-Physical Systems, February 25, 2014
12. Coleman, J.W., Malmos, A.K., Larsen, P.G., Peleska, J., Hains, R.: COMPASS tool vision for a system of systems collaborative development environment. In: International Conference on System of Systems Engineering, 2012 7th International Conference on, pp. 451–456, 16–19 (2012)
13. Agarwal, S., Ganguli, R.: Refining automated modeling of operational data by identifying the most important input factors. Min. Eng. **63**(12), 52–54 (2011)
14. Pape, L., Agarwal, S., Giammarco, K., Dagli, C.: Fuzzy optimization of acknowledged system of systems meta-architectures for agent based modeling of development. Procedia Comput. Sci. **28**, 404–411 (2014)
15. Konur, D., Dagli, C. (2014). Military system of systems architecting with individual system contracts. In: Optimization Letters, pp. 1–19
16. Agarwal, S., Pape, L.E., Dagli, C.H.: A hybrid genetic algorithm and particle swarm optimization with type-2 fuzzy sets for generating systems of systems architectures. Procedia Comput. Sci. **36**(133), 57–64 (2014)
17. Pape, L., Giammarco, K., Colombi, J., Dagli, C., Kilicay-Ergin, N., Rebovich, G.: A fuzzy evaluation method for system of systems meta-architectures. Procedia Comput. Sci. **16**, 245–254 (2013)
18. Ergin, N.K.: Improving collaboration in search and rescue system of systems. Procedia Comput. Sci. **36**, 13–20 (2014)
19. Wang, R., Dagli, C.H.: Executable system architecting using systems modeling language in conjunction with colored Petri nets in a model-driven systems development process. Syst. Eng. **14**(4), 383–409 (2011)
20. Wang, R., Agarwal, S., Dagli, C.: Executable system of systems architecture using OPM in conjunction with colored Petri Net: a module for flexible intelligent and learning architectures for system of systems. In: Europe Middle East & Africa Systems Engineering Conference (EMEASEC) (2014)
21. Agarwal, S., Wang, R., Dagli, C.: FILA-SoS, executable architectures using cuckoo search optimization coupled with OPM and CPN-A module: a new meta-architecture model for FILA-SoS, France. In: Boulanger, F., Krob, D., Morel, G., Roussel (eds.) Jean-Claude Complex Systems Design and Management (CSD&M), pp. 175–192. Springer International Publishing (2015)
22. Agarwal, S., Pape, L.E., Dagli, C.H., Ergin, N.K., Enke, D., Gosavi, A., Qin, R., Konur, D., Wang, R., Gottapu, R.D.: Flexible intelligent learning architectures for SoS (FILA-SoS): architectural evolution in systems of systems. Procedia Comput. Sci. **44**(2015), 76–85 (2015)
23. Schutze, O., Lara, A., Coello, C.A.: On the influence of the number of objectives on the hardness of a multi objective optimization problem. Evol. Comput. IEEE Trans. **15**(4), 444–455 (2011)
24. Ishibuchi, H., Tsukamoto, N., Nojima, Y.: Evolutionary many-objective optimization: a short review. In: IEEE Congress on Evolutionary Computation, pp. 2419–2426, June 2008

25. Coello, C.A.C.: An updated survey of evolutionary multi objective optimization techniques: state of the art and future trends. In: CEC 99. Proceedings of the 1999 Congress on Evolutionary Computation, 1999, vol. 1. IEEE (1999)
26. Agarwal, S.: Computational intelligence based complex adaptive system-of-system architecture evolution strategy. In: Ph.D. Dissertation, Missouri University of Science and Technology, May 2015
27. Singh, A., Dagli, C.H.: "Computing with words" to support multi-criteria decision-making during conceptual design. In: Systems Research Forum, pp. 85–99 (2010)

How Do Architects Think? A Game Based Microworld for Elucidating Dynamic Decision-Making

Johan de Heer

Abstract How do we think? A puzzling question given that humans may employ various actions, tactics and strategies during complex decision making tasks. Not to mention the influence of personality, style and intentions on judgment and decision-making. In this paper we focus on modeling Dynamic Decision Making (DDM) by utilizing actual in-game observations. We developed a 'game based microworld' through which we can capture and analyze players' reasoning behaviors. The use case is a bid for a complex system, in which we are interested in the contractor architects' DDM. Further, we explore various methods for game analytics that can be used to understand human reasoning. We conclude with several applications where game analytics may be utilized such as knowledge engineering, business intelligence, and training.

1 Introduction

Dynamic decision-making (DDM) is interdependent decision-making that takes place in an environment that changes over time either due to the previous actions of the decision maker or due to events that are outside of the control of the decision maker [1]. Dynamic decisions, unlike single choice decisions, are typically more multifaceted and occur in a certain time frame. Note, that even relatively simple choice behaviors are prone to several cognitive biases but are considered equally important regarding understanding judgment and decision-making [2]. DDM is daily practice for lots of people, such as complex systems' architects, working in professional industrial environments. Take, for example, a bid or tender, in which often a limited number of pre-selected contractors are asked to bid. Several challenges may happen during bid period, and the architect from the contractor side will normally deal with clarifications

J. de Heer (✉)
Thales Research and Technology/T-Xchange Netherlands, Twente University, De Horst (Building 20—WH226), Drienerlolaan 5, 7522 NB Enschede, The Netherlands
e-mail: johan.deheer@txchange.nl

G. Auvray et al. (eds.), *Complex Systems Design & Management*,
DOI 10.1007/978-3-319-26109-6_10

or corrections with the bid team concerning the offering. A bid team may consist of several experts with various roles viz, sales manager, system engineer, proposal writer, capture team leader, subcontractor, portfolio manager, pricing coordinator, etc. Typically, the architect needs to make judgment and decisions over time with respect to multiple criteria orientations such as, customer satisfaction, technology utilization and product portfolio enhancement. At the end of the bid period, the contractor compiles a bid for tender by a closing date and time: thus, time pressure plays an important role as well. Elucidating the dynamics of human reasoning and behaviors in these types of use cases, however, is an enigma and requires a theory of mind, appropriate theoretical concepts, methods and techniques for studying DDM.

This paper sketches a '*game-based-micro-world*' for studying Dynamic Decision Making. We illustrate this microworld with an example that enables us to study how architects make decisions during bid period. Further, we outline certain type of behavioral models that can be provided by game analytic services. Next, we hypothesize how these methods of 'stealth assessment' [3] can be applied for areas such as knowledge engineering, business intelligence and training.

2 Microworlds to Study DDM

Computer simulations are used to study Dynamic Decision-Making (DDM). These computer simulations are also named "microworlds" [4] and are used to observe and study how people make decisions over time. DDM differentiates from more classical forms of decision making by taken into account [2]: sequences of decisions to reach a goal, interdependence of decisions on previous decisions, dynamics of a changing environment, and that decisions are made in real time (time pressured situations). Microworlds become the laboratory analogues for real-life situations and aid DDM investigators to study decision-making by compressing time and space while maintaining experimental control [2]. This is important since empirical studies with a high degree of experimental control have the advantage of high levels of validity ('measure what we think we are measuring') and causality ('changing a variable here causes an hypothesized effect there'), however, at the expense of generalizability ('results found here count for lots of other situations as well'). Note however that even though microworlds aim to characterize the important elements of the real world, they differ in many respects from naturalistic decision-making (NDM)—low validity, low causality, high generalizability. Microworlds and DDM are somewhere in between highly controlled laboratory settings and real world situations.

3 Game Based Micro World

We designed and developed a game based microworld (see Fig. 1) that represents the essential real world elements during bid period from the architects' point of view. Note, that it is beyond the scope of this paper to discuss how we designed and

Fig. 1 Single player turn taking 2D narrative game

developed this model-based and configurable game based microworld. It needs understanding of designing game based systems [5], and a thorough understanding of the components that game systems are made of [6].

The game flow of this single player turn-taking narrative game based microworld is as follows. First, the architect—the player—is presented a context scenario, in which a bid context is briefly explained (Fig. 2).

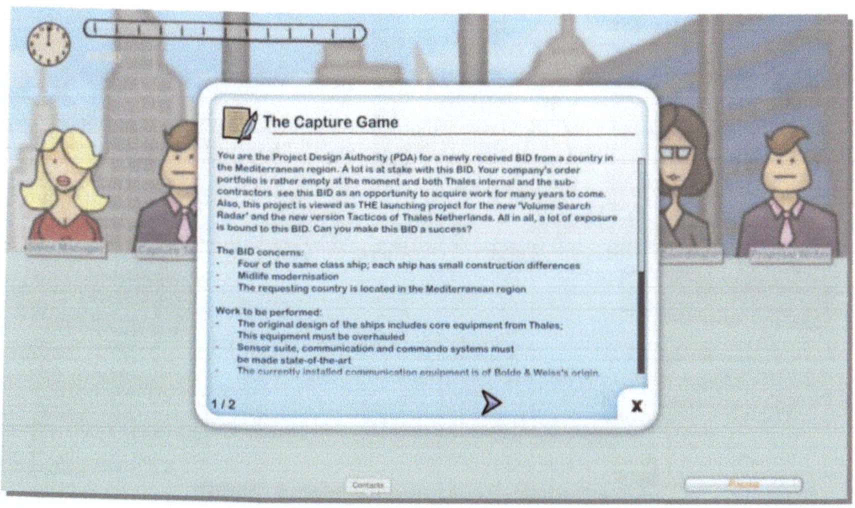

Fig. 2 Context scenario

Related to this context scenario, a series of dilemmas are introduced that end with a question where the player has to make a 'yes' or 'no' decision. The dilemmas—depicted in bottom left corner in the form of envelopes that can be opened with a simple mouse click—appear over the course of (playing) time. A typical scenario is the upgrade of several systems on a ship. An example of a dilemma in this scenario is: 'While you are exploring technical possibilities, a colleague points out the option of including a newly developed system type in the Bid. This holds the potential to improve the performance of the ships' upgrade. Question: Will you include the new sensor in the Bid? Note, that this game based microworld embeds dilemmas where

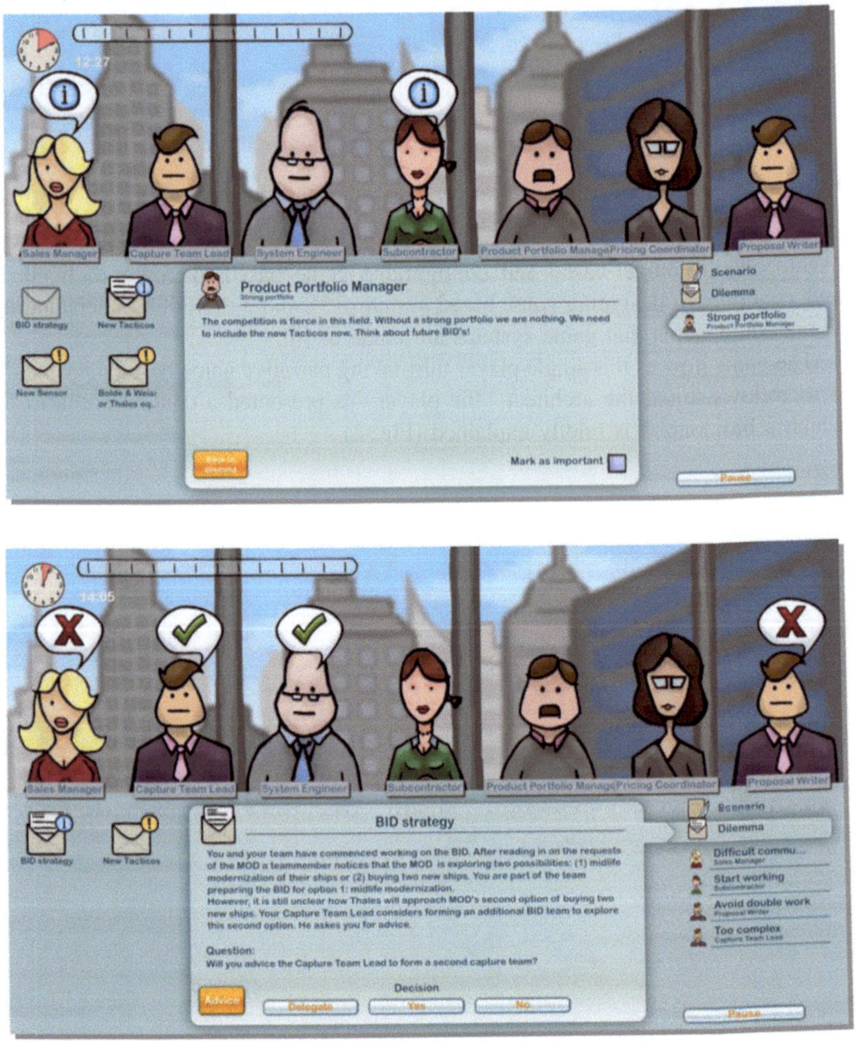

Fig. 3 Asking information and/or advice

there is no right or wrong answer. In the game (virtual) bid team members are gathered around a table and may let the player know that they have potential relevant information (depicted by a text balloon above their heads) that may possibly alter the decision if taken into account by the player. The player is free to select and read information from his team advisors, and may even ask them for advice what they would decide (Fig. 3).

Once, the dilemma has been answered, the game pauses and the player is firstly asked to indicate, which information provided by a virtual team member was taken into account and considered relevant regarding the decision he took. Virtual

Fig. 4 In-situ feedback

characters start to smile if the player 'listens' to them, but will look sad if players just 'hear' what they have to say. Secondly, he needs to indicate his perception with respect to the effect of his decision on the customer (see Fig. 4). After he provides this in situ feedback, the player returns to the game.

The game ends when all dilemmas have been answered or the time limit has been reached (15 min). The player may read all dilemmas first before answering any of them. The player is free to choose which dilemma he will answer first. The player may even delegate the dilemma if he thinks that it is not his responsibility to answer —of course, he will loose playing time if it is his responsibility. Besides that, extra information and even advice what to decide from team members is available per dilemma, but again, it is up to the player to decide if and when he uses this information. In other words, the player has several degrees of freedom in a rather constrained environment. A typical experiment with this game based microworld takes about 60 min. It starts with a 5 min introduction on the goals of the experiment. Then several game scenarios are played, which will take about 15 min for each scenario. The remaining time is spent on discussing in hindsight their thoughts and considerations while reaching their decision.

4 Game Analytics

Player behavioral modeling deals with the generation of models of player behavior. In general, a player model is an abstracted description of a (human) player's behavior in a game environment. Currently, we focus on player behavior modeling established via indirect measures of human players, by utilizing actual in-game observations. In future experiments we will also use direct-measurement approaches that make use of biometric data such as heart rate, galvanic skin response, EEG, etc.

The current version of the game based microworld generates the following player models via game analytic services. First, we generate simple descriptive statistics about the time needed to answer dilemmas, the number of dilemmas answered, the number of times advices of various team members were indicated as important (Fig. 5).

Second, we generate a newspaper article where the narrative (basically, the story the player tells) is based on the choices made during gameplay (Fig. 6).

Third (see Fig. 7), player' decisions are related to three different leadership dimensions, (1) 'Technology utilization', (2) 'Customer satisfaction', and (3) 'Product portfolio enhancement'. A graph shows how their answers (in %) are related to each of three dimensions.

Informally interviewing players after game play revealed that each of them inclined to have a preferred playing style during game flow, which suggested stable actions, tactics, and maybe even strategies for managing the uncertainty and dynamics in the game. The emergence of these stable behaviors could provide an opportunity for learning about human action, tactics and strategies vis à vis DDM

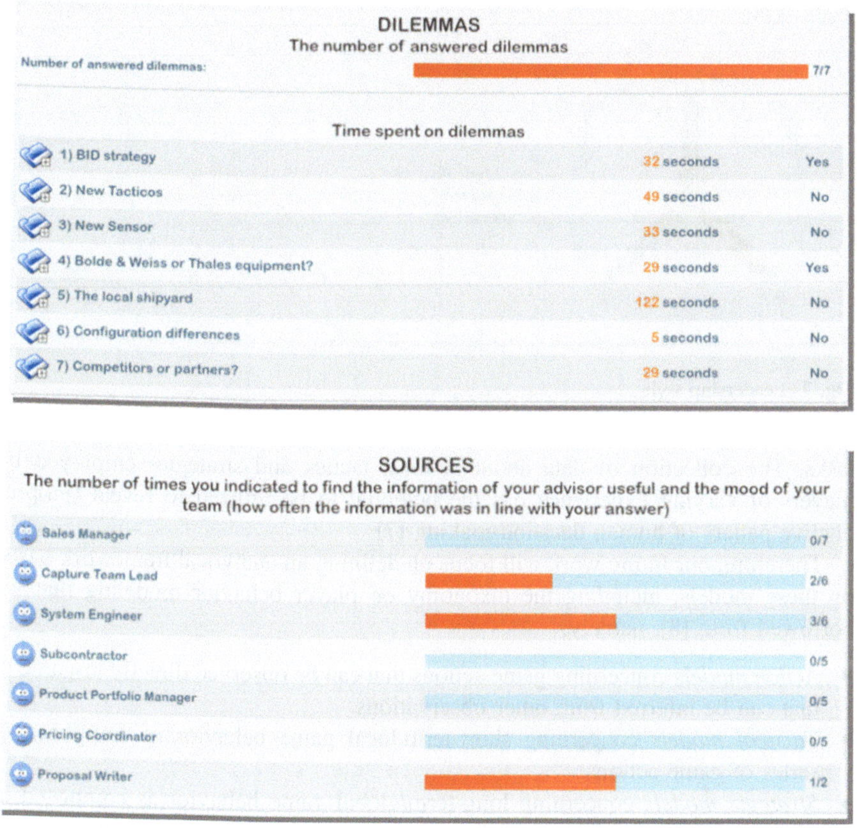

Fig. 5 Game statistics

NL NIEUWS

NIEUWS OPINIE BLOGS CULTUUR SPORT ECONOMIE REIZEN

Wednesday 18 February 2015

Reflections on succesfull BID

As previously reported, Thales won a BID for midlife modernization of four Mediterranean ships. The celebration have passed and now is the time to reflect on the mixed performance of the BID. Was the customer central in this BID? Will it create revenue or go over budget due to technology risks? And does the project show enough vision for the future of Thales products?

With this BID Thales lacks ambition to stay the world leader in combat management systems with their Tacticos product line. The new Tacticos version EDC +6 will not be included in this BID in order to reduce project risks.

At first glance, the admiral was not amused. However, he agreed with Thales argumentation that the strict time schedule was of more importance.

The MOD expressed challenging performance requirements in the BID regarding the upgrade of sensor suites of their ships. This introduced an interesting challenge to the BID. Thales is developing a promising new sensor suite, but the TRL has only just reached level 7.

The BID team successfully chose the sensible option. Work with the old model and provide our competent sales manager with a challenge suited for his skills.

Thales was at the verge of losing the BID due to a conflict over choice of communication equipment. Despite MOD's clear preference to Bolde & Weiss, the BID-team was urged to include the Thales communication equipment.

Including the Thales equipment proved a sensitive topic and during contract negotiations Thales had to take her losses on this issue. Thales was forced to revise their conditions and had to accept Bolde & Weiss as sub-contractor for the communication equipment in this BID. The business line in question was greatly disappointed over the loss of their position in the BID.

The local culture was underestimated by the Thales BID team. Especially the importance of family ties was not recognized. By not choosing the local shipyard the support of a key decision maker was endangered. Only the good friendship between Thales sales manager and the admiral saved the BID. The question is wether or not future oppertunities have been jeapordized with this lack of political insight.

The bid process was somewhat disturbed by a late proposal of a german consortium. Thales judged the situation well and remained confident in her BID.

Fig. 6 Newspaper article

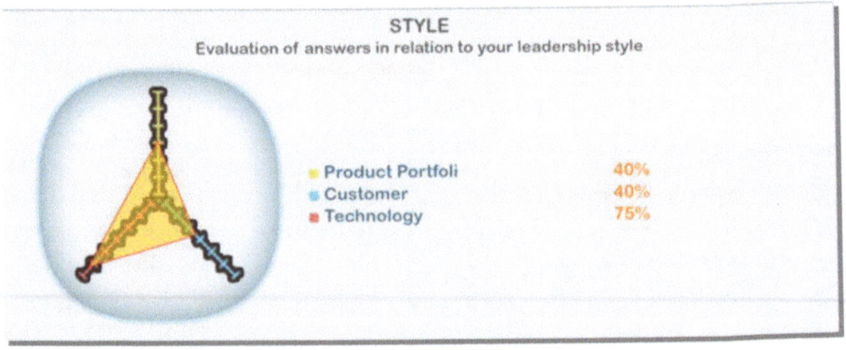

Fig. 7 Leadership style

tasks. The collection of data about actions, tactics and strategies employed by players of varying experience has the potential to be utilized to reveal (un)productive patters of human decision-making [7].

Therefore, our future work will focus on defining an analytical framework based on these models, including the taxonomy on player behavior modeling that we borrowed from [8], namely:

- *Action models* concerning game actions that can be observed directly in-game or that can be inferred from other observations.
- *Tactical models* concerning short-term/local game behavior as composed of series of game actions.
- *Strategic models* concerning long-term/global game behavior as composed of series of game tactics, of which behavior may span the entire game, several game iterations, and across distinct games.
- *Player profiling models* concerning the (psychological/sociological) profile of a player; this is important since distinct motivations and affect may result in distinct strategies, tactics and actions.

A method to implement player modeling is by modeling the actions that a player performs. A technique that has been proposed for building action models is sequential prediction, specifically by the use of N-grams [8]. N-grams are sequences of actions, e.g., selecting a dilemma, asking information from person y, asking advice from person z, making a choice, etc. If there are many actions to choose from then it might be hard to predict actions based on N-grams. In this case player models can be generated by probabilistic A* path analysis and consist of a probability distribution over all possible actions that the player might perform [8]. In addition, tactics are one to several orchestrated actions to realize a certain local goal. Tactics can be understood as the underlying drivers for player actions. Furthermore, strategies are concerned with the overall plan for reaching a long-term outcome. Likewise, modeling player strategies builds upon models of player tactics. As [8] points out: determining a player's local goal (tactics) generally requires fewer observations than determining a player's higher-order goal (strategy). Finally, player profile models tries to model internal traits of the player (e.g. personality and preferences) [8].

In order to facilitate more data collection on player behaviors, our game based microworlds are able to transform game data loggings on-the-fly to statistical packages that will further analyze these data streams into meaningful information to elucidate human reasoning. Not only individual behaviors but examination of team and group behaviors across a number of other parameters as well e.g. level of expertise, gender, country, culture, business line, etc. That activity is still underway and experiments conducted and data gathered will be addressed in future papers.

5 Conclusion

Following a focus on player behavioral modeling for Dynamic Decision Making (DDM) established via game based microworlds, in this article we distinguished several types of player models that we currently generate, and briefly discussed (1) action models, (2), tactical models, (3) strategic models, and (4) player profiles models. When considering the predictive capabilities of these types of models, most game developers would utilize these models to optimize the game experience. The general goal of player behavioral modeling often is to steer the game towards a predictable higher player satisfaction on the basis of modeled behavior of the human player [8]. Next to adaptive game systems, here we argue that these models have a lot of potential 'outside' the game based microworld. Inherently, behavioral models such as the ones we currently capture, further enhanced with action, tactical, strategic and player profiler models may help us to understand human reasoning in a more profound way.

We foresee various types of applications, for example, game based learning and assessments, game based business intelligence, and game based knowledge engineering. In game based learning these behavioral models can be used for in- and after-game feedback towards players for educational and training purposes. Configuring the game (in design- or run-time) to a player's behavioral model for reaching a particular learning objective. Through, for example, inquiry learning [9] players are encouraged trying out different playing styles in terms of actions, tactics and strategies. One of the game based training examples we developed is the so-called Mayor Game [10]. Dutch mayors are trained via this game regarding their leadership style during crises management. In this case—given the implemented didactical and instruction philosophy—trainees are stimulated to reflect and share their elucidated behavioral models. An example of game based business intelligence is the Cyber Security awareness game that we developed with a petro chemical game scenario. This game is facilitated by an organization that builds up customer intelligence models, and, therefore, let security managers from critical infrastructure organizations (their customers) play these games to understand their thinking in terms of their risk taking and—avoidance behaviors. Yet, another example is game based knowledge engineering. The idea here is that we let domain experts play a game to understand their reasoning strategies: how they make sense out of various forms of (un)structured data/information elements that pop-up during game play. The focus here is to elucidate how domain experts fuse data and information and reason with those information elements. For example, we developed a game that that shed light on the differences between experts and novices in terms of several cognitive biases and heuristics in forensics. This type of human reasoning is important for developing probabilistic reasoning—or decision support techniques for decision support systems. Therefore, we conclude that game based microworlds will bring us analytics in understanding how we think, reason, and decide that can be used for several purposes outside the game.

Acknowledgments Thales' Key Technology Domain Systems group for supporting this research & technology activity, Thales Netherlands Top Class Architecting for providing the trainees for this experiment, Thales Netherlands Naval Systems for development of the game content for the Bid scenario. T-Xchange, a research collaboration on serious gaming between Thales Research & Technology and Twente University that developed the game based microworld. COMMIT/who provided us the means and opportunity to set up a use case for game based knowledge engineering.

References

1. Gonzalez, C., Lerch, J.F., Lebiere, C.: Instance-based learning in dynamic decision making. Cogn. Sci. **27**, 591–635 (2003)
2. Kahneman, D.: Thinking, Fast and Slow. Farrar, Straus and Giroux, New York (2011)
3. Mayer, I.S., van Dierendonck, D., van Ruijven, T. et al.: Stealth assessment of teams in a digital game environment. In: GALA 2013 Conference, Paris, 23–25 October, pp. 1–13 (2013)
4. Brehmer, B., Dorner, D.: Experiments with computer simulated microworlds: escaping both the narrow straits of the laboratory and the deep blue sea of the fields study. Comput. Hum. Behav. **9**, 171–184 (2003)
5. Klabbers, H.G.: The Magic Circle: Principles of Gaming and Simulation, 3rd & rev. edn. SensePublishers, Dordrecht (2009)
6. Schell, J.: In Peters, A.K. (ed.) The art of game design: a book of lenses, 2nd edn. CRC Press, Boca Raton (2014)
7. Ross, A.M., Fitzgerald, M.E., Rhodes, D.H.: Game-based learning for system engineering concepts. In: Conference on Systems Engineering Research (CSER), pp. 1–11 (2014)
8. Bakkes, S.C.J., Spronck, P.H.M., van Lankveld, G.: Player behavioural modelling for video games. Entertainment Comput. **3**(3), 71–79 (2012)
9. de Jong, T., Linn, M.C., Zacharia, Z.C.: Physical and virtual laboratories in science and engineering education. Science **340**, 305–308 (2013)
10. van den Ven, J.G.M., Stubbé, H., Hrehovcsik, M.: Gaming for policy makers: it's serious! In: GALA2013, LNCS 8605, pp. 376–382 (2014)

EMI: Engineering and Management Integrator

Michael Masin, Yael Dubinsky, Michal Iluz, Evgeny Shindin and Abraham Shtub

Abstract The impact of systems engineering on program cost has been recognized for over a decade. From the very early stages, careful management of the relationships between the product design and the project plan is crucial to the success of any project that aims to deliver a defined product. Failure to closely manage the intricate web of resource constraints emanating from the two domains, the project scope and the product scope may lead to inadequate product performance or overruns in project schedule and budget. Identifying and managing the relationship between these two domains are at the heart of our challenge to combine project management (PM) and systems engineering (SE). We present a new approach, called EMI, which integrates SE and PM methodologies. These include the EMI mathematical foundation, implementation in architectural optimization and project management tools, and a detailed use case for development of the Doors Management System for commercial aircraft.

M. Masin (✉) · Y. Dubinsky · E. Shindin
IBM Research—Haifa, Mount Carmel, Haifa 31905, Israel
e-mail: michaelm@il.ibm.com

Y. Dubinsky
e-mail: dubinsky@il.ibm.com

E. Shindin
e-mail: evgensh@il.ibm.com

M. Iluz · A. Shtub
Technion—Israel Institute of Technology, Technion City, Haifa 34000, Israel
e-mail: iluzmichal@gmail.com

A. Shtub
e-mail: shtub@ie.technion.ac.il

G. Auvray et al. (eds.), *Complex Systems Design & Management*,
DOI 10.1007/978-3-319-26109-6_11

1 Introduction

"All programs…shall apply a robust Systems Engineering approach that balances total system performance and total ownership costs…" [1]. Meeting the required system specifications is one of the main challenges for systems engineers, whether chief system engineers or system engineers of a specific discipline. One of their objectives is to meet budget and schedule goals for the development and manufacture of the system and its subsystems. The ultimate goal is to provide a high quality product on time and within budget. Systems engineering (SE) and project management (PM) are two tightly intertwined domains, as stated in the Handbook of Systems Engineering and Management [2]. In contemporary practice, system engineers relate to the development and manufacturing costs that can be measured in work hours. However, when considering technological approaches, system engineers may choose state-of-the-art solutions without fully considering schedule and budget implications and constraints. In other words, they sometimes focus more on the product scope and neglect the project scope. System engineers make product-domain decisions that directly influence the project-domain, in which the project manager is responsible for multi-criteria decision-making (MCDM) on a daily basis. Analytic and simulation decision support tools can be of great value in such an environment.

Starting with the pivotal work of Forsberg and Mooz [3], many papers describe the need or conceptual framework for integrating SE and PM (see, e.g., [4–8]). Our work shows how this integration can be accomplished. This paper presents EMI, short for Engineering and Management Integrator, which combines SE and PM methodologies. Our contribution is in developing a practical mathematical model that combines SE architectural optimization and PM planning tasks. As part of our work, we implement this method using decision support tools, define a holistic engineering and management methodology, and demonstrate it for a typical product development.

The paper is organized as follows. We first describe the EMI mathematical foundation and methodology. We then demonstrate the methodology in a typical industrial use case involving the development of the Doors Management System (DMS) in a commercial aircraft using two state-of-the-art tools. The first tool is the Architectural Optimization Workbench (AOW), which deals with the architectural optimization (AO) aspects. The second is the Project Team Builder (PTB), which deals with the PM aspects. We show the limitations that arise when each tool is used on its own and how using EMI to integrate these tools can overcome the limitations and improve the results for finding an efficient DMS architecture. Finally, we summarize and provide directions for future research.

2 EMI Mathematical Foundation and Methodology

2.1 *Project Time Management for Architectural Optimization*

While PM has many aspects [9], in developing EMI we focus on the integration of time management and selection of the best system architecture. The mathematical model closest to the settings of AO is the Multi-mode Resource Constrained Project Scheduling Problem (MRCPSP) [10, 11]. In MRCPSP, project activities have several operational modes, each with its own duration and required set of resources. The activities have precedence constraints, and resources have a final capacity. An MRCPSP solution defines the mode in which each activity is executed and schedules the activities according to precedence and resource constraints. Solution procedures for MRCPSP include both heuristic and exact methods. In off the shelf MILP solvers, such as Cplex [12], the exact methods based on MILP formulation are capable of solving industrial size problems [13]. It is interesting to note, that the classical time-indexed formulation or its slight modification are usually the preferable options for solving real size problems with a few hundreds of periods [11].

We begin our methodology by adjusting MRCPSP to our AO needs. First and foremost, we synchronize the mode selection with the selected architectural components. Development projects are usually performed with preemptive schedules and part-time job intensity. This is especially true in matrix organizations, where technical units with domain expertise provide services to all running projects. Our model also supports variable period lengths. The total number of periods and consequently, model size, could be significantly reduced using longer period lengths for later periods, where having a detailed plan makes less sense. The model we developed, called AO-MRCPSP, is described below.

Sets: $\boldsymbol{A}$—activities, $\boldsymbol{IP_i}$—immediate predecessors of activity $i \in A$, $\boldsymbol{M_i}$—modes of activity $i \in A$, R—resources, $\boldsymbol{P} = 1\ldots\boldsymbol{T}$—periods, $\boldsymbol{G}$—subsystem/components types. **Parameters**: $\boldsymbol{w}$—the minimum work intensity if an activity is performed in a part-time, $\boldsymbol{e}$—maximal extension of activity duration caused by preemption, $\boldsymbol{p_t}$—duration of period $t \in P$, $\boldsymbol{T_{max}} = \sum_{t \in P} p_t$—time horizon, $\boldsymbol{d_j}$—duration of mode j, $\boldsymbol{r_{jk}}$—requirement of mode j for resource k, $\boldsymbol{a_j}$—resource independent cost of mode j, $\boldsymbol{h_i}$—subsystem/component type of activity i, $\boldsymbol{n_j}$—subsystem/component type id of mode j, $\boldsymbol{v_k}$—capacity of resource k, $\boldsymbol{b_k}$—cost of resource k per time horizon. **Decision variables**: $\boldsymbol{x_{ij}}$—binary variable for mapping activity $i \in A$ to mode $j \in M_i$, $\boldsymbol{y_{jt}}$—continous variable for mapping mode j to period $t \in P$, $\tilde{\boldsymbol{y}}_{it}$—binary indicator that activity $i \in A$ is performed at period $t \in P$, $\boldsymbol{s_{it}}$—binary wave variable that activity $i \in A$ started at period $t \in P$ or earlier, $\boldsymbol{f_{it}}$—binary wave variable that activity $i \in A$ finished before period $t \in P$ or earlier, $\boldsymbol{C_i}$—completion time of activity $i \in A$, $\boldsymbol{C_{max}}$—completion time of the whole project, $\boldsymbol{u_{kt}}$—utilization of resource k at period t, $\boldsymbol{u_k}$—average utilization of resource k, $\boldsymbol{A}$—total cost of activities, $\boldsymbol{B}$—total cost of resources, $\boldsymbol{D}$—total cost of project, $\boldsymbol{q_o}$—type id of subsystem/component type $o \in G$.

$$\textbf{AO} - \textbf{MRCPSP} \quad \text{Minimize}\{C_{max}, D\} \\ \text{Subject to} \tag{1}$$

$$\sum_{j \in M_i} x_{ij} = 1 \quad \forall i \in A \tag{2}$$

$$y_{jt} \leq x_{ij} \quad \forall i \in A, j \in M_i, t \in P \tag{3}$$

$$w \cdot \tilde{y}_{it} \leq \sum_{j \in M_i} y_{jt} \quad \forall i \in A, t \in P \tag{4}$$

$$\tilde{y}_{it} \geq \sum_{j \in M_i} y_{jt} \quad \forall i \in A, t \in P \tag{5}$$

$$s_{it} \leq s_{i,t+1} \quad \forall i \in A, t \in P | t < T \tag{6}$$

$$f_{it} \leq f_{i,t+1} \quad \forall i \in A, t \in P | t < T \tag{7}$$

$$s_{it} \geq \tilde{y}_{it} \quad \forall i \in A, t \in P \tag{8}$$

$$f_{it} \leq 1 - \tilde{y}_{it} \quad \forall i \in A, t \in P \tag{9}$$

$$\tilde{y}_{it} \leq f_{i't} \quad \forall i \in A, i' \in IP_i, t \in P \tag{10}$$

$$\sum_{j \in M_i} \sum_{t \in P} y_{jt} p_t \geq \sum_{j \in M_i} x_{ij} d_j \quad \forall i \in A \tag{11}$$

$$\sum_{t \in P} (s_{it} - f_{it}) p_t \leq e \sum_{j \in M_i} x_{ij} d_j \quad \forall i \in A \tag{12}$$

$$C_i = T_{max} - \sum_{t \in P} f_t p_t \quad \forall i \in A \tag{13}$$

$$C_{max} \geq C_i \quad \forall i \in A \tag{14}$$

$$u_{kt} = \frac{1}{v_k} \sum_{i \in A} \sum_{j \in M_i} y_{jt} r_{jk} \quad \forall k \in R, t \in P \tag{15}$$

$$u_{kt} \leq 1 \quad \forall k \in R, t \in P \tag{16}$$

$$u_k = \frac{\sum_{t \in P} u_{kt} p_t}{\sum_{t \in P} p_t} \tag{17}$$

$$B = \sum_{k \in R} u_k v_k b_k \tag{18}$$

$$A = \sum_{i \in A} \sum_{j \in M_i} x_{ij} a_j \tag{19}$$

$$D = A + B \tag{20}$$

$$\sum_{j \in M_i} x_{ij} n_j = q_{h_i} \quad \forall i \in A \tag{21}$$

$$x_{ij}, \tilde{y}_{it}, s_{it}, f_{it} \in \{0, 1\} \quad 0 \leq y_{jt}, u_{kt}, u_k \leq 1 \quad C_i, C_{max}, A, B, D \geq 0 \tag{22}$$

The total project duration and cost are common objective functions in Eq. (1). Moreover, any piecewise linear function of decision variables could be added to the set, for example, the utilization range of critical resources to smooth their usage. Constraints (2)–(3) ensure exactly one mode for each activity. Constraints (4) define minimal part-time intensity w if activity i is performed during period t, for example, 50 % to allow working on two activities at most. Note, only one y_{jt} can be positive. Constraints (5) connect mode continuous performance to an activity binary indicator at period t. Constraints (6)–(9) ensure correct behavior of wave functions when activity i starts and finishes (because of objectives and other constraints, s_{it} and f_{it} try to become 1 as late and as early as possible, respectively). Constraints (10) ensure precedence relation, allowing activity i to start only after all its predecessors have finished. This constraint could be relaxed to $\tilde{y}_{it} \leq f_{i',t+1}$ to allow simultaneous execution of an activity with its predecessors for rough resource allocation (e.g., in later long periods). Constraints (11) ensure activity i is performed long enough to be completed. Effectively, these constraints imply that the project should be completed before T_{max}. Constraints (12) restrict the total time activity i has started but not finished yet to e times its nominal duration d_j. Constraints (13)–(14) calculate activity and project completion time, respectively. Activity start time could be calculated similarly using s_{it} instead of f_{it}. Additional constraints could be added to the earliest and latest start and completion times. Constraints (15)–(17) calculate resource unitization and constrain it according to the available capacity. We allow variable resource capacity over periods in our implementation. Constraints (18)–(20) calculate the cost of resources, B, the cost of activities, A, and the total project cost, D, respectively. Constraints (21) relate the PM model with AO, ensuring that activities connected to some subsystem/component choose the same subsystem/component type as AO. Constraints (21) define domain of decision variables. AO-MRCPSP can incorporate time and resource buffers. Dummy activities without required resources (i.e., $r_{ik} = 0$ for all k) before integrating activities and before activities that request critical resources represent time and resource buffers, respectively.

The AO-MRCPSP formulation above has several beneficial properties. Allowing preemption and part-time intensity help relax regular MRCPSP constraints and remove the need for binary indicators per mode per period. In this setting, event-based formulations [14], usually most applicable for large projects, are less attractive since preemptions require more events. In addition, the model is not very sensitive to the number of modes—most variables and constraint sets are per activity, not per mode. Probably the most important property of AO-MRCPSP is that it requires a relatively small number of periods, which is the most sensitive and problematic parameter of time-indexed formulation. Currently, standard optimization packages, such as Cplex [12] can handle up to a few hundreds of periods. If each period represents a week, our model optimizes a multi-year plan. During AO, a rough estimation is required with a lot of uncertainty regarding later stages of the project. The period lengths could be adjusted accordingly, further reducing the model size. For example, in our use case we apply bi-weekly periods.

2.2 AO-MRCPSP in Architectural Optimization

To incorporate the aspect of architectural optimization into EMI, we relied on two preliminary works. The first is the *concise modeling* [15] extension of SysML [16] to specify architectural alternatives and system constraints. Concise modeling combines regular SysML diagrams that define architectural topology (e.g., components multiplicity and connections rules), with associated data tables called *catalogs* for block subclasses and *inventories* for part multiplicities. Using concise modeling, AO assumes that all parts are *a priori* optional, and divides all attributes into either *parameters* determined by data tables or SysML model values, or *variables* optimized during AO. For example, when outlined by *catalog* stereotype, the RDC block has an associated catalog table listing several RDC options and their relevant parameters (weight, power, etc.). These options could include different technologies, such as optical or copper cables, that should be synchronized with the PM choices of appropriate activities in their modes. The optimization process is responsible for finding concrete architecture alternatives that conform to the topology, driven also by a predefined set of system or subsystem attributes called *design objectives*, and constraints grouped by different types of analysis, called *analysis viewpoints*. Analysis viewpoints help calculate the system variables or define their feasible region for architectural optimization, considering concerns such as cost, weight, reliability, timing, resource allocation, and power and data distribution. The second preliminary work [17] defines the concept of *pluggable analysis viewpoints*, demonstrating its ability to specify design objectives as a library of reusable assets. The pluggable viewpoints are based on the concept of *classification-by-property* [18], in which the computational semantics for the sets appearing in constraints is specified according to different properties of the domain elements. Masin et al. [17] demonstrate that applying classification-by-property principles creates robust analysis libraries that remain resilient to system evolution during product design, and enables Lego type interoperability between different system analyses. An adaptation of AO to PM is straightforward using the methodology for pluggable analysis viewpoints: the AO-PM formulation below just adds "analysis viewpoint" AO-MRCPSP to other viewpoints, where constraints (25) are similar to (21) applied to subsystem/component catalogs.

$$\textbf{AO}-\textbf{PM}\quad \text{Minimize}\ \textit{Original objectives},\ \textit{Objectives}(1) \quad \text{Subject to} \tag{23}$$

$$\textit{Original architectural constraints} \quad \text{Constraints}(2)-(22) \tag{24}$$

$$\textit{Subsystem/component type synchronization constraints} \tag{25}$$

2.3 EMI Methodology

EMI methodology focuses on the integration of System Architecture Synthesis with Project Time Management. In most MBSE and PM methodologies, both tasks are relatively independent, performed by different people using different tools. EMI defines what information is transferred between the tools to implement the AO-PM model and obtain a holistic system architecture and project plan.

The initial information comes from the SE team to define system-related data such as subsystems, subsystem options, and performance goals. Then, the PM team defines project activities including precedence between activities, alternative modes, and required resources. The PM data is incorporated with other viewpoints considered during the design space exploration (DSE) for MCDM. The selected Pareto optimal solutions are transferred to the PM team for detailed analysis. If the results are satisfactory, the project can start. Otherwise, the PM team should update project parameters, especially the adjusted resources capacities, activity and resource buffers, and mode durations. New architectures are then found by the AO-PM model. The whole process is shown in Fig. 1. EMI can be used as a pre-project fuzzy front-end stage [19] in the early planning stages of the project or during the project. Each usage requires slight modifications in the suggested

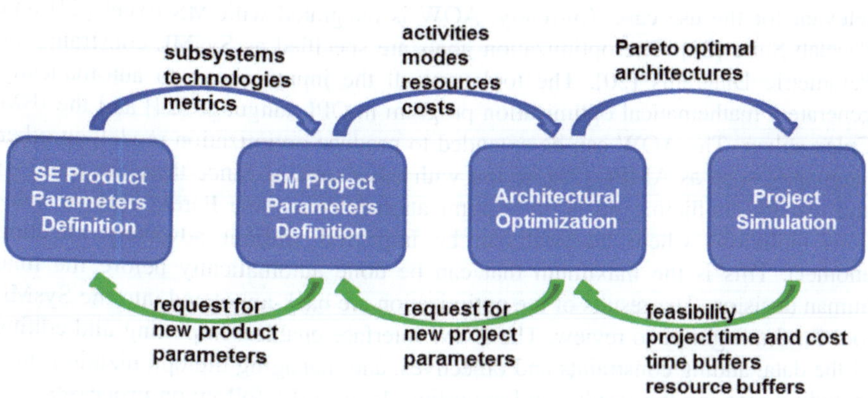

Fig. 1 EMI process

process, for example, customer interaction in fuzzy front-end or re-iteration to address system and environment changes during the project. Although we distinguish between SE and PM, in practice, the teams could be interdisciplinary.

3 Demonstration of EMI Methodology

3.1 SE and PM Tools

In this section we briefly describe two state-of-the-art tools for SE and PM that we use to demonstrate the EMI methodology. System engineers are typically responsible for creating alternative architectural solutions according to all requirements and goals, and for choosing the best one. However, the ever-increasing complexity of systems, strict design constraints, conflicting goals, and many other factors make the process of finding optimal designs extremely difficult. The common means to achieve system goals is to build optimization models for the set of specified architectures and find the best one using suitable optimization software tools. Unfortunately, this approach is highly labor-intensive because each architecture solution created by the engineer requires a separate optimization model created by an expert. This issue is resolved by the **Architectural Optimization Workbench** (AOW) described in [15, 17], and [20]. In the AOW, the system engineer can rapidly create the necessary system architecture, satisfying all functional and technical constraints needed to achieve the specified goals. Using standard SysML with the *concise profile* described above, the system engineer can model the composition rules (also known as architectural patterns, or templates) of the required functional and physical system structures and relations inside (data flow, energy flow, etc.) and between them (potential mapping between functions and physical components). In Rational Rhapsody [21] this approach immediately allows linking the functional models to the requirements in DOORS. The potential physical components are imported from a library, along with geometrical data, if relevant for the use case. Currently, AOW is integrated with MS Excel [22] and Pacelab Suite [23]. The optimization goals are specified as SysML constraints or Parametric Diagrams [20]. The tool uses all the inputs above to automatically generate a mathematical optimization program in OPL language [24] and the IBM Cplex solver. The AOW can be extended to produce optimization models in other languages, such as AMPL [25], to use with other solvers. Since there are multiple and usually conflicting goals, the optimization finds diverse Pareto optimal solutions (solutions where no goal can be improved without adversely affecting another). This is the maximum that can be done automatically before the final human decision. The results of the optimization are back-annotated into the SysML tool for the engineer to review. The AOW interface enables: importing and editing of the data, adding constraints and objectives, and managing the optimization runs, including viewing the results, and exporting them to the follow-on processes.

Our second tool, the **Project Team Builder** (PTB) is an integrated decision support system designed to support new product development teams during the project [26–32]. PTB combines simulation and case study approaches. Each case study is a new product development project performed under schedule, budget, and resource constraints, in a dynamic stochastic environment. The details of these case studies are built into the simulation, and all the data required for analysis and decision-making is easily accessed by the user interface. Random effects simulate the uncertainty in the environment, and decisions made by the user cause changes in the state of the simulated system. PTB supports a model-based approach for translating the voice of the customer into the project scope. A database is built into the PTB to support decision-making, post factum analysis, and backtracking. A friendly GUI enables a typical user to learn how to use the PTB within an hour. The PTB combines classical PM domains such as scheduling of activities with management of requirements. It offers a module for managing system requirements that supports the process of selecting alternative designs to determine system performance. The simulator allows the generation of project scenarios that include stochastic activity duration, resource capacity, and costs. Based on the input, the simulator offers a forecast for the project cost, schedule, and the product quality.

3.2 *EMI Use Case with Doors Management System*

To evaluate the benefits of our EMI methodology, we applied it to a use case involving the Doors Management System (DMS) for an aircraft. The DMS controls the latching and locking of doors in an aircraft. It communicates with the aircraft's pressurized system and consists of sensors, actuators, controllers, data collectors, and an Avionics Full-Duplex Switched Ethernet (AFDX) network. Functional requirements define the system functions, such as sensing, latching, locking and controlling, and the data flow between them. The structure describes potential DMS topologies, and the geometry gives potential physical locations. Figure 2 shows the transition of a customer story to a *concise model* in which all requirements and design alternatives are formally defined. The system optimization criteria include system weight, cost, and power consumption. Marketing requires project completion within one year (48 weeks).

As a comparison to EMI, we applied the SE methodology to the design of DMS for commercial aircraft. The AOW takes into account the system design for the aircraft door sensors, remote data concentrators, controllers, actuators and related power cables, power and data distribution flow, and safety requirements while optimizing the results for weight, cost, and power consumption. The AOW found four Pareto optimal solutions, shown in Fig. 3a. Systems engineers prefer to focus on two solutions with weight less than 200 kg, and power less than 305 W.

We also applied the PM methodology to the use case using PTB. The project activity network for the DMS use case is shown in Fig. 3b. Project simulations run in PTB showed that none of the two architectures coming from the SE team could

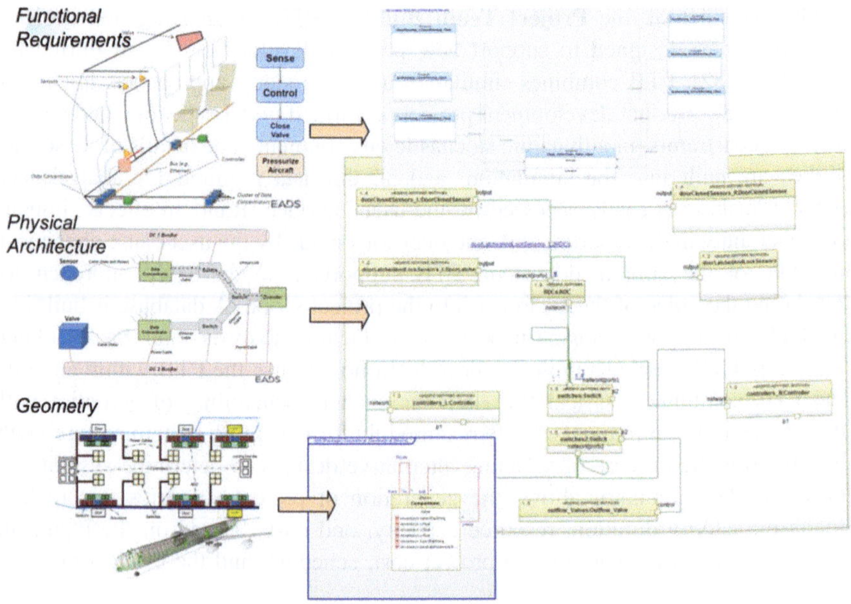

Fig. 2 From customer's story to *concise modeling*

be developed in 48 weeks. By defining the benefit as a weighted sum of the weight, material cost, and power (with 25, 50, and 25 %, respectively), the PM team could approximate their values based on the available activity modes, and they found a completely different Pareto frontier, shown in Fig. 3c. Unfortunately, all the chosen architectures resulted in weight and power above the required threshold set by the SE team. In the next section, we applied EMI methodology, integrating both tools, AOW and PTB, for the DMS use case.

Using the EMI methodology for the same use case, we implemented AO-PM in AOW and provided data communication between AOW and PTB. The resulting Pareto frontier from AOW contains two solutions, as shown in Fig. 4a. Both were not on the original Pareto frontiers obtained by AOW and PTB. The first architecture transferred to PTB and passed all simulation tests (Fig. 4b) and was chosen for DMS development. Compared with architectures shown in Fig. 3, the chosen system has better chances for success since both the SE and PM teams have not compromised their objectives and constraints, and together found a well-balanced design architecture with a manageable design process.

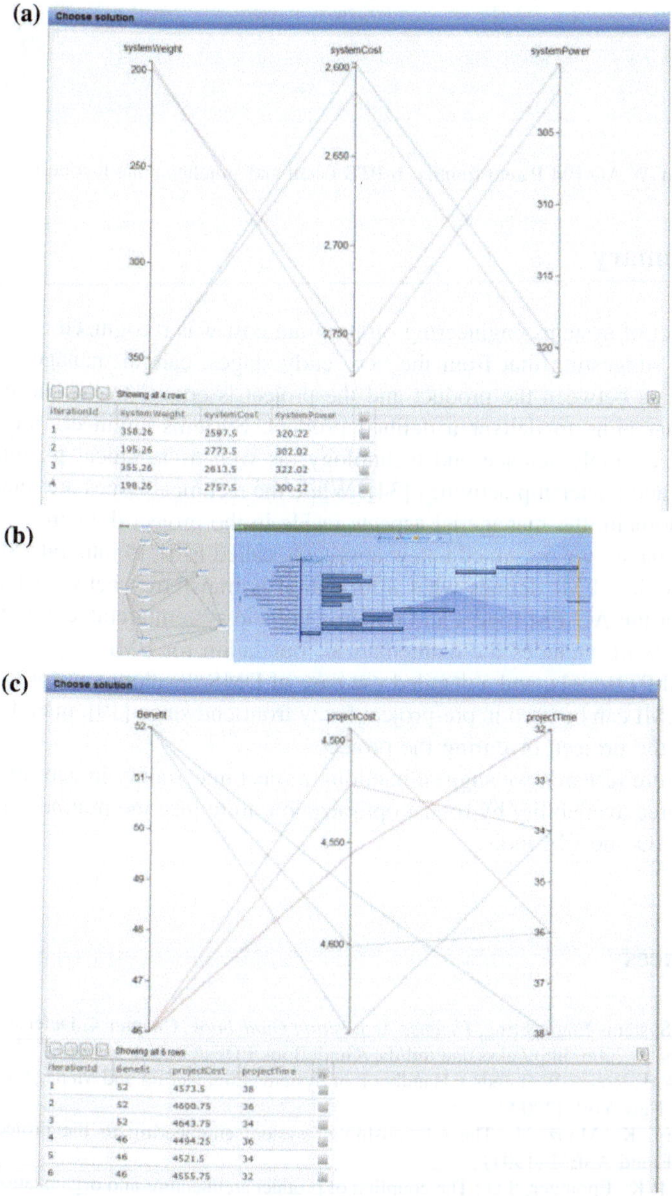

Fig. 3 **a** AOW only Pareto frontier. **b** DMS development activity precedence diagram and Gantt chart. **c** PTB only Pareto frontier

(a)

iterationI...	systemW...	systemC...	systemP...	projectTi...	projectC...
1	195.3	2775.5	302.02	46	4662.5
2	195.3	2775.5	302.02	48	4635.25

(b)

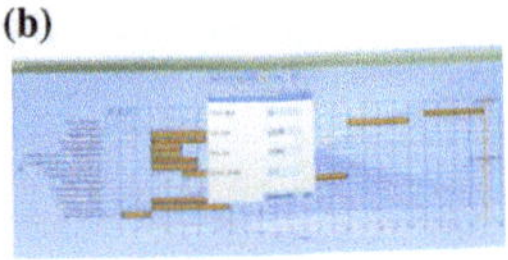

Fig. 4 **a** AOW AO-PM Pareto frontier. **b** PTB Gantt and simulation the first design

4 Summary

The impact of systems engineering on program cost was recognized over a decade ago [33], suggesting that from the very early stages, careful management of the relationships between the product and the project is crucial to the success of any project that aims to deliver a defined product. Systems engineers are required, therefore, to apply science and technology, as well as technical planning, management, and leadership activities [34]. While the technical issues are related to the product domain, the managerial aspects reside in the project domain.

In this paper, we presented a new approach, called EMI, to integrate SE and PM methodologies. EMI defines what information is transferred between the tools to implement the AO-PM model and obtain a holistic system architecture and project plan. Our work includes the mathematical foundation for EMI, implementation in AOW and PTB tools, and a detailed use case of DMS development for commercial aircraft. EMI can be used in pre-project fuzzy front-end stage [19], in early planning stages of the project, or during the project.

For future research we suggest handling project uncertainty in activity durations and resource availability by robust optimization, to reduce the number of iterations between AO and PM tools.

References

1. DAU, Systems Engineering, *Defense Acquisition Guidebook*, Chapter 4, Defense Acquisition University, 2006, http://akss.dau.mil/dag/GuideBook/PDFs/Chapter_4.pdf
2. Sage, A.P., Rouse, W.B. (eds.): Handbook of Systems Engineering and Management, 2nd edn. Wiley, New York (2009)
3. Forsberg, K., Mooz, H.: The relationship of system engineering to the project cycle. In: NCOSE and ASEM (1991)
4. Gulati, R.K., Eppinger, S.D.: The coupling of product architecture and organizational structure decisions. In: Working Paper. MIT (1996)
5. Sosa, M.E., Eppinger, S.D., Rowles, C.M.: Identifying modular and integrative systems and their impact on design team interactions. In: Jof Mechanical Design (2003)
6. Cataldo, M., Herbsleb, J.D., Carley, K.M.: Socio-technical congruence: a framework for assessing the impact of technical and work dependencies on software development productivity. In: 2nd ACM-IEEE International Symposium on ESEM (2008)
7. MacCormack, A., Baldwin, C., Rusnak, J.: Exploring the duality between product and organizational architectures. Res. Policy (2012)

8. Avnet, M.S.: Information flow, team coordination, and shared knowledge in integrated concurrent engineering. In: INCOSE (2014)
9. Project Management Institute: *PMBOK® Guide,* 5th edn (2013)
10. Schwindt, C., Zimmermann, J. (eds.): Handbook on Project Management and Scheduling, Volume 1, Springer (2015)
11. Zapata, J.C., Hodge, B.M., Reklaitis, G.V.: The multimode resource constrained multiproject scheduling problem: alternative formulations. AIChE J. (2008)
12. Cplex Optimizer. www.ibm.com/software/commerce/optimization/cplex-optimizer/
13. Artigues, C.: Recent developments in mixed integer linear programming formulations for the resource-constrained project scheduling problem. PMS (2014)
14. Koné, O., Artigues, C., Lopez, P., Mongeau, M.: Event-based MILP models for resource-constrained project scheduling problems. In: C&OR (2011)
15. Broodney, H., Dotan, D., Greenberg, L., Masin, M.: Generic approach for systems design optimization in MBSE. In: INCOSE (2012)
16. OMG Systems Modeling Language. http://www.omgsysml.org/
17. Masin, M., Limonad, L., Sela, A., Boaz, D., Greenberg, L., Mashkif, N., Rinat, R.: Pluggable analysis viewpoints for design space exploration. In: CSER (2013)
18. Parsons, J., Wand, Y.: Emancipating instances from the tyranny of classes in information modeling. ACM Trans. Database Syst. **25**(2) (2000)
19. Katz G.: Rethinking the product development funnel (2011)
20. Masin, M., Broodney, H., Brown, C., Limonad, L., Mashkif, N., Sela, A.: Reusable derivation of operational metrics for architectural optimization. In: CSER (2014)
21. Rational® Rhapsody. http://www-01.ibm.com/software/awdtools/rhapsody/
22. MS Excel. https://products.office.com/en-us/excel
23. Pace Lab Suite. http://www.pace.de/products/preliminary-design/pacelab-suite.html
24. Van Hentenryck, P.: The OPL optimization programming language (1999)
25. AMPL. http://www.ampl.com/
26. Parush, A., Davidovitz, L., Shtub, A.: Simulation-based Learning in engineering education: performance and transfer in learning project management. JoEE (2006)
27. Davidovitch, L., Parush, A., Shtub, A.: Simulation-based learning: the learning-forgetting-relearning process and impact of learning history. C&E (2008)
28. Shtub, A., Parush, A., Hewett, T.: Guest editorial: the use of simulation in learning and teaching. IJEE (2009)
29. Parush, A., Davidovitch, L., Shtub, A.: Simulator-based team training to share re-sources in a matrix structure organization. IEEE Trans. EM (2010)
30. Shtub, A.: Simulation based training (SBT)—the next generation of project management training. PM World J. (2013)
31. Shtub, A., Iluz, M., Gersing, K., Oehman, J., Dubinsky, Y.: Implementation of lean engineering through simulation based training. PM World J. (2014)
32. Cohen, I., Iluz, M., Shtub, A.: A simulation-based approach in support of project management training for systems engineers. Syst. Eng. (2014)
33. Defense Systems Management College: Systems Engineering Fundamentals (1999)
34. Frank, M.: Cognitive and personality characteristics of successful systems engineers. INCOSE (2000)

Property Model Methodology: A First Application to an Operational Project in the Space Domain

Erwann Poupart, Jean-Marie Wallut and Patrice Micouin

Abstract The purpose of this paper is to provide a feedback on a Model Based Systems Engineering application to a space domain project. In the core of the paper, and after a synthetic presentation of the systems engineering methodology called Property Model Methodology (PMM), the case study, coming from the space domain, is described. In this context, PMM has been used in order to validate a top-level textual specification and to define the verification scenarios and verification cases aiming at establishing the correctness and the completeness of the physical system developed according to this top-level textual specification. The paper provides first feedbacks about PMM utilization. The conclusion summarizes the benefits and also the limitations that are identified today, and includes a presentation of the future works.

1 Introduction

There is a general agreement on the idea that there is a crisis of the classical systems engineering [1], as well as there was a software engineering crisis starting from the ninety's. Whether we consider the domains of energy production, of transport vehicle industry (road, rail or air) and even in space industry, the manifestations of this crisis are still the same: delivery delays, objective cost overruns and lack of

E. Poupart (✉) · J.-M. Wallut
CNES Centre Spatial de Toulouse, 18 Avenue Edouard Belin, 31401, Toulouse Cedex 9, France
e-mail: erwann.poupart@cnes.fr

J.-M. Wallut
e-mail: jean-marie.wallut@cnes.fr

P. Micouin
Arts et Métiers ParisTech LSIS UMR CNRS, 2 Cours Des Arts et Métiers, 7296, 13617 Aix-En-Provence, France
e-mail: patrice.micouin@incose.org

G. Auvray et al. (eds.), *Complex Systems Design & Management*,
DOI 10.1007/978-3-319-26109-6_12

maturity of the systems put into service. If the causes are certainly many, one of them is the growing gap between the means used (a widely document-centric engineering) and, on the one hand, growing objectives assigned to systems under development and, on the other hand, the conditions in which these systems are developed (large teams, from various geographical, linguistic and, cultural areas). From this generally shared understanding, the proposals for resolving this crisis diverge. For pragmatists, it consists of detecting the minimum corrective actions to obtain the greatest improvements, such as the establishment of good practice guides. Others will look into more agile methods in order to reduce misunderstandings that abound in development teams. We could designate them as inter-subjectivists ("*people rather than processes*[1]"). Finally, the last one, see a solution in a strengthening of formality and rigor of the implemented processes and exchanged products. If, in our opinion, each of the above approaches has a grain of truth and deserves to be explored, we side clearly with the last one: we assume that the crisis of the classical systems engineering, of which principles were designed decades ago, can find a solution thanks to formality in the development processes, and the exchange, between stakeholders, of engineering products as little interpretable as possible and as accurate as possible, starting with validated specification models and verified design models. In what follows, we present PMM method, we provide a report of its use in the context of a space project and we make a feedback on lessons learnt.

2 PMM: Goals, Processes and Concepts

PMM is a recently developed Model-Based Systems Engineering methodology [2]. Its compound name comes from two of its main characteristics, (1) the formulation of requirements based on the concept of property (property-based requirements or PBR) and (2) the adoption of a model-based systems engineering (MBSE) approach. This is a very classical descending development approach; however it authorizes the reuse of preexisting blocks. This development approach is compatible with current industrial development standards, specifically ARP4754A, EIA632 or Space Engineering Standards. It is also built on a third pillar, namely simulation, which is the primary means for validating specification models and verifying design models, while the verification of physical products, their integration and installation are maintained (Fig. 1).

Roughly described, PMM may start just after the validation of a Concept of Operations (CONOPS) [3] and is made up of the following activities (obviously, it includes a lot of recovery points to reengineer the system, when goals are not the right ones or are not reached):

[1]Agile Manifesto: http://www.agilemanifesto.org/sign/display.cgi?ms=000000309.

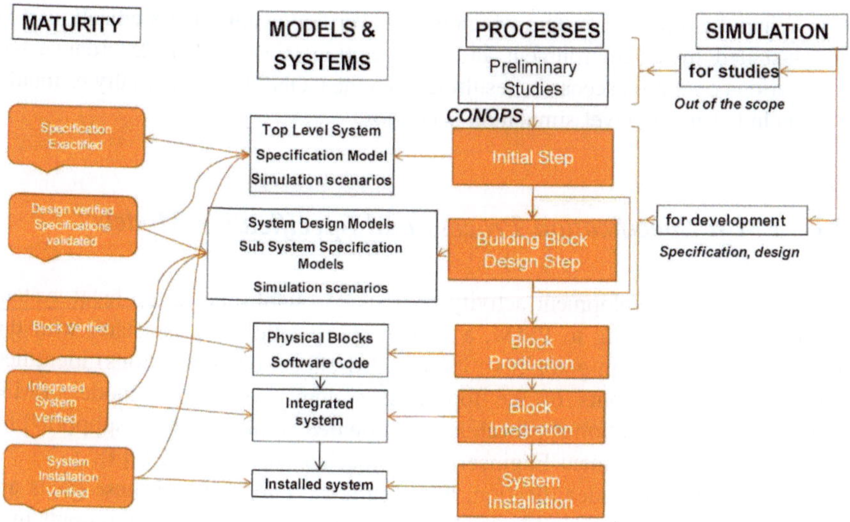

Fig. 1 PMM process model

(1) The initial step includes the establishment of the top level system specification model and the elaboration of simulation scenarios built in order to validate the top level system specification model. The system model is ready for validation. When this goal is reached, the top level system specification model may be considered as exact (as exact as possible).

(2) Subsequent steps are a finite repetition of a building block design step including the establishment of a system design model providing us with an architecture of the building block (alternate architectures may be considered and a preferred one may be selected). When this architecture introduces subsystems, equipment, items or parts, a specification model for each of these components is established, while the connections among these components are stated. Simulation scenarios are built of each of these component specification models. When this goal is reached, the low level subsystem specification models are validated step by step against the system specification model. The design model of a building block is not further decomposed when its behavior may be directly formalized as equations or when it may be picked up from a building blocks catalogue.

(3) The design process ends when all the elementary building blocks are designed or acquired. Then, step by step, elementary building block design models, and integrated building block design models are verified against their own specification models.

(4) Production and verification processes of elementary physical building blocks, their physical integration into intermediary building blocks and the associated verifications are described in [2, Chap. 11].

(5) When the top level system which is fully verified against its specification, it is installed in its environment and its operation is verified in accordance the various operation scenarios resulting from the CONOPS and partially or totally included in top level simulation scenarios.

2.1 PMM Specification Process and Specification Models

The first system development activity consists of establishing a top level system specification. According to PMM, a system specification process starts with the definition of the system goals (i.e., its intended effects or its functions) identified and modeled as outputs of the top level system specification model. PMM requirement determination approach is goal oriented, similar to those supported by Goal Oriented Requirement Engineering approaches such as KAOS [4].

Based on this goal identification, the occurrence conditions of these goals are elicited. During this process, expected inputs are identified while other outputs may be also identified and modeled, such as observable states, undesired effects and system failures. Undesired inputs may be also considered.

Then, we formalize the result of this elicitation in the form of PBRs [2, chap 6]. These PBRs are predicates linking together goals, secondary outputs, observable states and inputs. They specify system properties and their conditions of actualization. In particular, assumptions are specific PBRs related to input properties since they are always out of the system's developer control and only presumed (Fig. 2).

The basic form of a PBR is as follows:

$$PBR\text{: }[when\ Condition\ =\ >\ ]\ val(System.Property)\ \in\ Domain,$$

A conjunction operator "∧" of PBRs is defined allowing the combination of several PBRs while a partial order relationship "≤" allows to compare two PBRs. PBR1 ∧ PBR2 is the conjunction of two PBRs and is also a PBR while $PBR_1 \leq PBR_2$ means PBR_1 is less constraining than PBR_2.

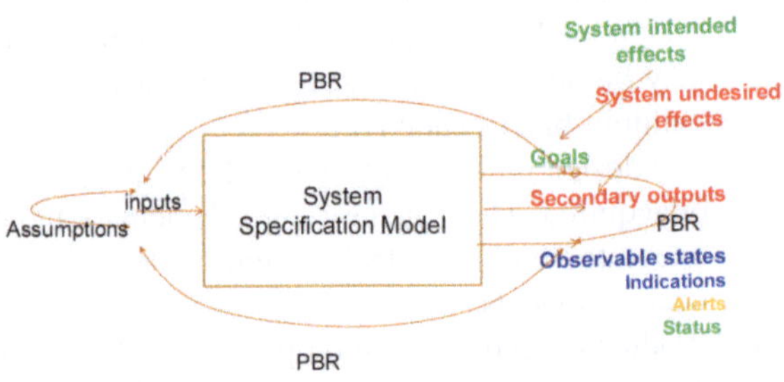

Fig. 2 Specification model

A specification model is a formal model dealing with (1) system requirements, (2) system interface requirements and (3) system assumptions. Translated in simulation models based on languages such as VHDL-AMS [5] or Modelica [6], a system model is syntactically coherent and complete. Simulation provides assistance for establishing the exactness of a system model for a defined set of validation scenarios.

2.2 *PMM Design Process, Design Models and System Verification*

When the exactness of a system model is established, the second system development activity is to conceive a system design model. This activity is based on the designer's knowledge of business rules, experience on the system domain and innovations. Several candidate architectures may be considered. Since the focus of this paper in not on design activities, they are not described here (they are detailed in [2, Chap. 7]).

For candidate architectures, the third activity is to derive the system requirements (PBR) into subsystem requirements {PBR_1, …, PBR_n}. To be valid, such derivations shall be such that, when the architecture A is selected, then the conjunction of {PBR_1, …, PBR_n} shall be more constraining than the system requirement PBR:

$$Derivation\colon when\, A\ =\ > PBR \leq PBR_1 \wedge \ldots \wedge PBR_n,$$

This validity condition entails the following theorem (called "*the prime contractor theorem*"): "*A sufficient condition for a system to comply with its PBRs is that its subsystems comply with the PBRs validly derived from the system PBRs, provided the design choices and assumptions made about the environment driving the derivation remain valid*".

Simulation provides, firstly, a means for establishing the validity of system PBRs derivation into a set of subsystem derived PBRs for a given level of rigor (related to the richness of validation scenarios). Secondly, simulation provides a means for verifying building block design models regarding their specification models.

During simulation sessions, the specification models monitor the corresponding design models, checking that for all submitted simulation scenario whether requirements are violated or not. When no requirement violation is detected for the complete verification of a building block design model, the building block design model may be considered as free of error regarding its specification model and for the considered effort of verification. And so on, up to the system level.

3 The Application: PEPS System Modeling

3.1 Description of the PEPS System

PEPS stands for Environmental Politics & Space Politics. It aims at developing usage of space images. This new CNES project, that has started its operational phase the 18th of June this year, is also part of Copernicus project (2014–2020) which is the most ambitious Earth observation program to date.

Copernicus is the new name for the Global Monitoring for Environment and Security program (GMES). This initiative is headed by the European Commission (EC) in partnership with the European Space Agency (ESA). ESA is developing a new family of satellites called sentinels that will provide a unique set of observations, starting with the all-weather, day and night radar images from Sentinel-1A, launched in April 2014. Sentinel-2 will deliver high-resolution optical images for land services and Sentinel-3 will provide data for services relevant to the ocean and land. Sentinel-4 and -5 will provide data for atmospheric composition monitoring from geostationary and polar orbits, respectively. Sentinel-6 will carry a radar altimeter to measure global sea-surface height, primarily for operational oceanography and for climate studies.

PEPS is the French ground system that will provide a public access to sentinels image products.

For the first time, multi-sensor, multi-scale and multi-temporal data over the Earth will be available for free which should help to create and develop environment services.

Users will be able to analyze data over a long period up to the beginning of the mission (climatic change: glacier footprint, ice field surface, lakes surface, desertification, vegetation indication), deforestation, urban expansion, road network, hydrology, volcanology, etc.

In 2017, PEPS system shall be able to store and provide access to 6.1 petabytes of data and 8 millions of products (image products, metadata and quick look). It shall also be extensible and able to store and provide access to 17 petabytes of data and 20 million of products in 2020.

After the end of 2017, PEPS system shall also provide, close to image products, high performance computing capability so that scientists or other external partners interested in developing geographic added-value services can run efficiently dedicated algorithms.

3.2 PMM Application Context and Goals

It is well known that requirements are the corner stone between specification/design, effective system (or implementation), and tests (verification that the system meets its requirements).

Previous R&T study at CNES (Limbes R&T project in 2008) has shown benefits to build a property centered design to ease system verification.

More than that, much more benefits can be obtained with higher quality of requirements in terms of semantic due to the many human actors involved in engineering requirement processes.

When system specification has been produced, reviewers will have to share a common understanding of requirements and produce RID's (Review Item Discrepancy) and this for each phase of the system engineering process:

- Phase A (Mission/operational analysis and feasibility):
  - Customer requirement review
- Phase B (Ground Segment Preliminary Design):
  - System requirements review
  - Preliminary design review
- Phase C (Ground Segment detailed design generally done by a sub-contractor):
  - Critical design review
- Phase D (Ground Segment production and verification):
  - Technical qualification review
  - Operational qualification review

Just after phase B and just before phase C, subcontractors involved in tender responses will also have to share a common understanding of requirements to design the most competitive solution.

In the same time, operational teams involved in the system as a part of it, will also have to design their tasks using the system to achieve mission goals.

Finally, requirements semantic will be checked once more during critical design review of phase C, and again to verify that the system including operations meets requirements during phase D reviews.

There is clearly a huge potential of improvement of engineering processes efficiency if requirements quality is improved (less ambiguity in semantic interpretation).

PMM provides a methodology and models to help system engineers to focus and make more explicit system goals. The main difficulty will be to integrate it successfully with the different human actors involved in the engineering process.

CNES had already made an experiment of PMM in the context of an R&T study in 2014 and developed a modelling front-end mock-up, named PMM Designer that was first experimented with a satellite imagery mission control system specification.

Even if the scope of the case study was limited in size, the result was that PMM concepts are robust to express requirements semantics, simple to use (due to its goal orientation, its concepts parsimony), and, even if the front-end mock-up could be enhanced, many potential benefits have been identified during this experiment, including:

- Improved quality and completeness of requirements,
- Efficiency of production and validation processes of requirements
- Efficiency of verification processes
- Efficiency of system maintenance in operational condition and its design

That were the reasons why we applied partially PMM to express PEPS system requirements more formally when PEPS system verification task started in March this year. At this time, PEPS specifications were already written in textual form in one document for all the releases from 1.1 until 1.4. They represent in total around 180 textual requirements.

The main goal was to assess benefits on our engineering process and limitations of the methodology and associated tool.

3.3 PBRs Determination and PBRs Validation

The first step was to create all the PBRs starting from PEPS system textual requirements release 1.1 (about 45 textual requirements for this 1.1 release). This formalization task was relatively efficient. Only two man weeks were necessary with the front-end mock-up to create a first release of 11 PBRs covering all the 45 requirements.

The Fig. 3 here after represents a snapshot of PMM Designer describing the PMM Top-Level Specification Model of PEPS acquisition System, with its expected outputs (goals), observables states and assumed inputs linked together thanks to PBRs (predicates) defining the outputs and their actualization conditions, assumptions on inputs, and so on. We started from 12 requirements and only 4 PBRs were sufficient to express PEPS acquisition system goals.

We observed that many textual requirements are very precise for the constraints about inputs but not correlated with the expected outputs and many missing constraints about outputs. We had to look at the implementation to characterize the intended effects.

Formalizing requirements into PBRs does gather execution context constraints with the expected outputs, so it gathers many textual requirements in one place.

Concretely, we have observed that 45 textual assertions, usually referred as textual requirements, were consolidated and restructured into 11 PBRs. Moreover, we significantly improved the quality of these requirements (providing rich, coherent and synthetic information contents).

Described in PBRs form, requirements are then verifiable, system can be tested, and moreover, other people responsible for designing and running the tests don't have to interpret again textual requirements. This requirement clarification task is done once and only once. We have already save some time to design the tests and we expect to save much more time in discussions and meetings usually spent by many people involved in the project trying to interpret textual requirements written by someone else.

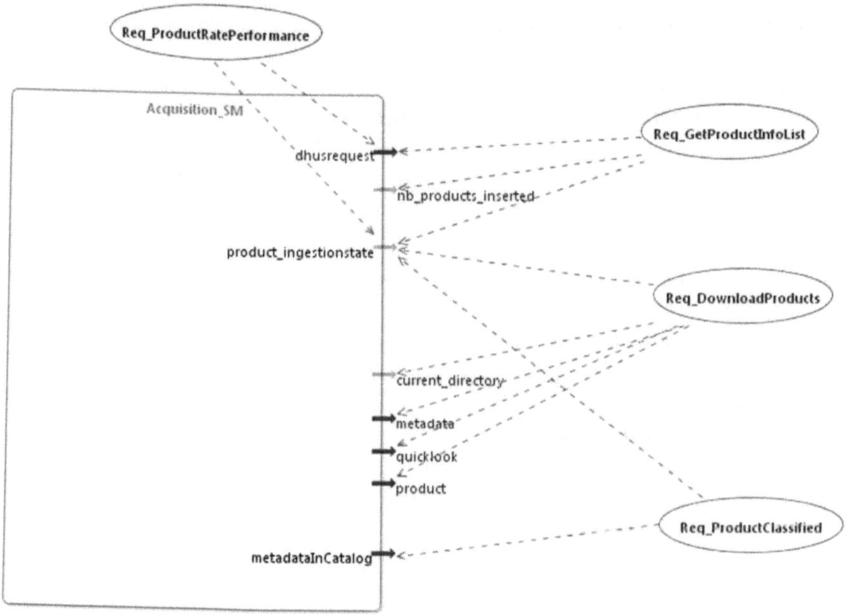

Fig. 3 PEPS acquisition system specification model with PMM designer

Another lesson learnt is related to security requirements; we had to take into account about 300 textual requirements. However, as there was already scripting tests that are able to check precisely those requirements, it was not useful to apply PMM to build the tests. We noticed that requirements were already unambiguous, for each requirement assertions about operating system properties are checked. They had spontaneously the structure of PBRs.

As a first example the PBR called "**Req_DownloadProducts**" in Fig. 4 specifies that, 12 h, at most, product items shall be complete on PEPS infrastructure and that we shall have an error to be handled by the human operator only if automatism cannot handle it.

Human operators are also goal driven and will compensate automatism when the goal is clearly defined and if he can have adequate observability of the system.

Note: Some logical operators are in comment because there are not available currently in PMM designer mock-up.

Writing PBR asks good questions to the specification author, it helps him or her to characterize a realistic and verifiable goal. For example, reading requirements ACQ-1810, ACQ-1015, ACQ-2030 we had to answer to the following questions: What is product item integrity? All product items shall be present or not? How can it be verified?

About the question: "*Where product shall be stored*"? We found 4 requirements (ACQ-2010, ACQ-2020, ACQ-2030 and AUT-0120) scattered in the document but only one PBR was sufficient using infrastructure architecture knowledge.

```
predicate Req_DownloadProducts(product_ingestionstate, current_directory, metadata,
quicklook, product):
    when (product_ingestionstate.downloadTaskStatus == "DOWNLOADING") delay 12 hours =>
        // ACQ-1015 => ACQ-2030 (name of product items)
        product.name == product_ingestionstate.productId//".zip" and
        metadata.name==product_ingestionstate.productId//"_Medata.xml" and
        quicklook.name==product_ingestionstate.productId//"_quicklook.gif"and
        // ACQ-1810 (product item integrity)
        //(product_ingestionstate.md5 == md5(product)) and
        // unzip -lz OK (readable zip) and
        // metadata starts with tag <product> and ends with tag </product>and
        // ACQ-2010, ACQ-2020, ACQ-2030 and AUT-0120 (product localization)
        ($current_directory == YYYY/MM/DD/SSS/) and (SSS == S1A or S1B or etc. satellite)
        // ACQ-0820 (failure handling)
        // if only uploading process running
        ((product_ingestionstate.downloadTaskStatus == "DONE") and
        (product_ingestionstate.ingestionTaskStatus == "TODO")
        or
        (product_ingestionstate.downloadTaskStatus == "ERROR") and
        (product_ingestionstate.ingestionTaskStatus == null))
```

Fig. 4 PEPS acquisition system PBR first example

We had also to identify missing observable state for the download task and to specify more precisely what to do in case of external site error (requirements ACQ-0820 that says "*After any failure, acquisition function shall resume from its current state.*"). If the error is only temporary, download task automatism shall retry, otherwise download task shall go to "error state" to be handle by human operator. The existing prototype implementation helped us to identify the missing observable states and intended effects.

This was not so easy to define because acquisition system is not able to distinguish surely between permanent or temporary errors because it does catch external site error messages that are not precise enough. PBR has to take into account what the system is able to perceive from the environment otherwise we will obtain an unreachable goal.

As a second example the PBR called "**Req_ProductClassified**" in Fig. 5 specifies that, 12 h, at most, product items shall be classified and stored in PEPS catalog infrastructure and that we shall have an error to be handled by the human operator only if automatism cannot handle it.

It shall be noted that there were **no requirements specifying this acquisition system goal about product classification**. This missing requirement has been identified during this PBR determination and validation phase. Product classification is an intended effect that required to be characterized.

```
predicate Req_ProductClassified(metadataInCatalog, product_ingestionstate):
when (product_ingestionstate.ingestionTaskStatus == "INGESTING") delay 12 hours =>
  // ACQ-0820 (failure handling)
  (product_ingestionstate.ingestionTaskStatus == "DONE") and
  // Missing classification product Requirement
  (metadataInCatalog contains product_ingestionstate.productId) and
  // landCover not empty and
  // geographical names not empty and
  // localization not empty and
  // id, spacecraft, characteristics not empty
or
  // ACQ-0820 (failure handling)
  (product_ingestionstate.ingestionTaskStatus == "ERROR") and
  not (metadataInCatalog contains product_ingestionstate.productId)
```

Fig. 5 PEPS acquisition system PBR second example

We obtained that product identifier shall be present in catalog database and characterized (land cover (water, forest, city, agriculture, etc. percentage); geographically localized (continent, country, region, city, etc.) instrument characteristics (instrument, processing level, product type, sensor mode, etc.), date, orbit number, satellite, resolution, snow cover, cloud cover if optical, etc.

This acquisition system goal is very important to characterize because it helps to check completeness and consistency of the peps images product catalog.

Note: For this first experimentation, only the presence in catalog was checked during the verification tests, assertions relative to product classification were human checked using Graphical User Interfaces -GUIs-. We can imagine checking it automatically in the future.

Then, the second step was to validate requirements formalization with its author to check that semantic interpretation was correct. The result is conclusive, only some minor points were reported. The author of the specification, who is a co-author of the present paper, agreed that this may help to improve our engineering processes even if it's easier for him to read the textual form of the PBR and not the predicates that are more adequate for computers

Then, those formalized requirements were presented to the contractor that will be responsible for the next releases of PEPS system from 1.2 until 1.4. Usually test design and specification task is done by the contractor that shall try to catch precisely requirements semantics and shall have some discussions with the author of the specification to clarify them. Once formalized, PBRs were presented to the contractor using PMM Designer. The result of this presentation was that the contractor agreed with the benefits of a more efficient mutual understanding of requirement semantics. They requested to use the PMM Designer as a possible support for this common understanding.

3.4 PEPS System Verification

For the verification tasks, it still remains to implement the test generally in a scripting language, even if we can imagine later some code generation starting from PMM models. PBR determination and validation facilitate test design task because what is to be checked is clearly defined.

We have translated PBRs into the scripting tests language (SQL requests and shell scripts). For the PBRs to be verified by humans (through GUIs), we translated them in human-readable statements. We can imagine automate some of those tests using other scripts in the future.

Technical qualification for PEPS system V1.1 was done successfully.

For the operational qualification, we had to show to human operators PEPS system goals using the application in release 1.1. We focused on expected outputs, observable states and system failures to be compensated by the operator. We observed that PBR's fits well with this operational process and we only had to show the real system's behavior when it is driven by all the PBR-based test scenarios and not the predicates and any requirements.

4 Conclusion: Lessons Learnt and Future Works

Benefits: the benefits observed in the experiment in 2014 are confirmed by this first application to the PEPS operational project:

- The requirements determination is goal-oriented, with as consequence, an improvement and a facilitation of requirements production,
- The quality and completeness of requirements are also significantly improved with a clear connection between expected outputs and the conditions of their actualization (observable states, failures, inputs).
- The efficiency of the validation process is also improved. Even inexact, a PBR remains unambiguous, measurable and testable. Its interpretation is done once and only once, preventing an undefined number of misinterpretations by the various stakeholders.
- Although it is still too early to claim it, the efficiency of the verification process should be significantly improved. When validated, PBRs are unambiguous (no possible misinterpretation) and testable, connected to their conditions of actualization as expected for test cases.

Limitations: Currently, the main limitations observed are related to the tool. PMM Designer is a preliminary mock-up of a modeling and simulation PMM tool. It provides the main editors requested by PMM. However, it does not provide a complete PBR editor, many operators are missing. It needs improvements to ease the modelling process. It does not interface either simulation back-end, thereby making it impossible to PBRs validation and design verification by simulation, while this validation / verification by simulation is one of the strengths of the method.

Future works: It is planned to iterate the process described in this paper for the next PEPS releases from 1.2 until 1.4. We will have of course many other lessons learnt to report by the end of this year with those coming releases.

References

1. Newport, J.R.: Avionic Systems Design. CRC Press (1994)
2. Micouin, P.: Model Based Systems Engineering: Fundamentals and Methods. Wiley & ISTE (2014)
3. IEEE Standard 1362, System definition—concept of operations document (1998)
4. von Lamsweerde, A.: Requirements Engineering. Wiley (2009)
5. IEEE Standard VHDL Analog and Mixed-Signal Extensions, IEEE 1076-1, IEEE Computer Society (2007)
6. Modelica Association, Modelica®—A unified object-oriented language for systems modeling language specification version 3.3 (2012)

A Model-Driven Approach to Enable the Distributed Simulation of Complex Systems

Paolo Bocciarelli, Andrea D'Ambrogio, Alberto Falcone, Alfredo Garro and Andrea Giglio

Abstract The increasing complexity of modern systems makes their design, development and operation extremely challenging and therefore new Systems Engineering and Modeling and Simulation (M&S) techniques, methods and tools are emerging, also to benefit from distributed simulation environments. In this context, one of the most mature tools is the IEEE 1516-2010—Standard for M&S High Level Architecture (HLA). However, building and maintaining distributed simulations components, based on the IEEE 1516-2010 standard, is still a challenging and costly task. To ease the development of full-fledged HLA-based simulations, the paper proposes the MONADS method that, according to the model-driven systems engineering paradigm, allows one to generate the HLA-based simulation code from SysML models by the use of a chain of *model-to-model* and *model-to-text* transformations. The effectiveness of the method is shown through a case study that concerns an Automated Transfer Vehicle (ATV) approaching and docking to the International Space Station (ISS).

Keywords Modeling and simulation · High level architecture · Model-driven systems engineering · Distributed simulation

A. Falcone · A. Garro
Department of Informatics, Modeling, Electronics and Systems Engineering, University of Calabria, Via P. Bucci 41C, 87036, Rende, Italy
e-mail: alberto.falcone@dimes.unical.it

A. Garro
e-mail: alfredo.garro@dimes.unical.it

P. Bocciarelli (✉) · A. D'Ambrogio · A. Giglio
Department of Enterprise Engineering, University of Rome "Tor Vergata", Via del Politecnico 1, 00133, Rome, Italy
e-mail: paolo.bocciarelli@uniroma2.it

A. D'Ambrogio
e-mail: dambro@uniroma2.it

A. Giglio
e-mail: andrea.giglio@uniroma2.it

G. Auvray et al. (eds.), *Complex Systems Design & Management*,
DOI 10.1007/978-3-319-26109-6_13

1 Introduction

Systems are constantly increasing in complexity and sophistication involving several heterogeneous components that are often designed and developed by organizations belonging to different engineering domains, including mechanical, electrical, and software. Moreover, moving from large-scale systems to Systems of Systems (SoSs), the involved components (that can be regarded as systems themselves) are often geographically distributed and capable of autonomous and independent behaviors. In addition, during the life of a SoS, as new systems may join the SoS and other dynamically may leave it, its components and their relationships typically change. This increasing level of complexity makes the design, development and operation of modern systems extremely challenging. As a consequence, new Systems Engineering methods and techniques are emerging also to benefit from Modeling and Simulation (M&S) distributed simulation environments [1].

In this context, the IEEE 1516-2010—High Level Architecture (HLA) standard [2]—supports the simulation of modern complex systems by providing a distributed infrastructure in which each simulation unit runs on an independent computer (in general, geographically distributed) and communicates with the others in a common simulation scenario. HLA was developed by the U.S. Modeling and Simulation Coordination Office (M&S CO) to facilitate the integration of distributed simulation models within a common architecture. Although it was initially developed for purely military applications, it has been widely used in non-military industry for its many advantages related to the interoperability and reusability of distributed simulation components. In the HLA standard a distributed simulation is called a *Federation* and it is composed of several HLA simulation entities, each called a *Federate*, which interact among them by using a Run-Time Infrastructure (RTI), the backbone of a Federation execution that provides a set of standard protocols and services to manage the communications and data exchange among Federates. Each Federation has a Federation Object Model (FOM) that is created in accordance with the Object Model Template (OMT) defined by the standard [2, 3]. A FOM contains specifications of *Object classes* (objects are instances—or entities—of object classes that have attributes that can be updated), *Interaction classes* (a message sent among objects that has parameters) and *Data types* (the technical specifications and semantics of attributes and parameters).

Building complex and large distributed simulations components, based on the IEEE 1516 standard, is usually a challenging task and requires considerable effort, not only in their development, but also for the cost of maintaining such components. On the development side, the building and testing of *HLA Federates* is generally difficult, complex, and resource-intensive because of the complexity of the IEEE 1516 standard [2, 3], the lack of proper documentation, and the availability of ready-to-use examples. Moreover, developers have to spend a considerable effort to face with common HLA aspects, such as the management of the simulation time, the connection on the HLA/RTI, and the management of common

RTI exceptions. As a result, they cannot fully focus on the specific aspects of their own *HLA Federates*.

To ease the development of full-fledged HLA-based simulations, model-driven software engineering (MDSE) approaches, tools and techniques could be effectively exploited. MDSE is an approach to software design and implementation that addresses the rising complexity of execution platforms by focusing on the use of formal models [4]. According to this paradigm, a software system is initially specified by the use of high-level models. Such models are then used to generate other models at a lower level of abstraction, which in turn are used to generate other models, until stepwise refined models can be made executable. One of the most important initiatives driven by MDSE principles is the Model Driven Architecture (MDA) [5], the OMG's (Object Management Group) incarnation of Model-Driven Engineering (MDE).

In this paper, MDA is exploited to design and develop the MONADS method (MOdel-driveN Architecture for Distributed Simulation) that aims at facilitating the distributed simulation of complex systems, specified by using SysML, according to the MDSE paradigm. Moreover, the HLA simulation code, generated starting from SysML models by a chain of *model-to-model* and *model-to-text* transformations, is based on the *HLA Development Kit software Framework (DKF)*, a software framework released under the open source policy Lesser GNU Public License (LGPL) by the University of Calabria, working in cooperation with the NASA JSC (Johnson Space Center), and that allows one to support the development of reliable HLA Federates by managing their lifecycle and handling the common HLA aspects.

The paper is structured as follows. In Sect. 2, the MDA- and DKF-based MONADS method is illustrated. The method is exemplified in Sect. 4 by considering the reference scenario described in Sect. 3 and that concerns a situation in which an Automated Transfer Vehicle (ATV) is approaching the International Space Station (ISS) to dock on it. Related proposals are discussed in Sect. 5; whereas, in Sect. 6, conclusions are drawn and future works delineated.

2 The MONADS Method

MDA-based software development is founded on the principle that a software system can be built by specifying a set of model transformations, which allow to obtain models at lower abstraction levels starting from models at higher abstraction levels.

To achieve such an objective, MDA has introduced a language for specifying technology neutral metamodels (or models used to describe other models), referred to as the Meta Object Facility (MOF) [6], and a standard for specifying model transformations, i.e., the Query/View/Transformation (QVT) standard [6]. A model transformation specified in QVT allows one to automatically generate a target model, instance of a given MOF-based metamodel, from a source model, instance of the same or of a different MOF-based metamodel. In case the target model is of text type (e.g., code written in a given programming language), the MOFM2T (MOF Model To Text) standard [6] can be used to specify the relevant transformation.

According to the context outlined in Sect. 1, the system development process is concerned with two different engineering domains. On the one hand, it is related to the system development domain, in which *systems engineers* deal with design and implementation issues. On the other hand, it addresses the simulation development domain, in which *simulation engineers* deal with system verification and validation issues by introducing distributed simulation-based analysis techniques. In this respect, the proposed method supports both system and simulation engineers, as depicted in Fig. 1.

At the beginning, the system under study is specified in terms of a SysML model (e.g., block definition diagrams, sequence diagrams, etc.). According to the MDA terminology [5], such a model is referred to as the *platform-independent model* (*PIM*) of the system. At the system development level, the system engineer in charge of producing the system model is not concerned with any details regarding the simulation model and is strictly focused on the specification of a SysML-based system design model, starting from the system requirements.

The SysML model identifies the input of the sub process that is related to the development of the distributed simulation. In this respect, according to the DSEEP

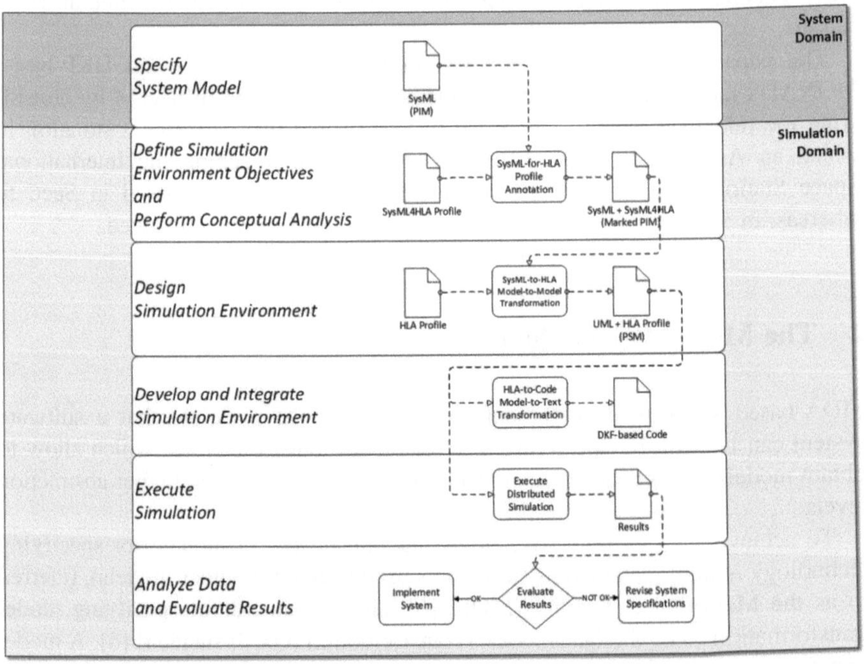

Fig. 1 Overview of the MONADS method

(Distributed Simulation Engineering and Execution Process) standard [7], simulation engineers carry out a *conceptual analysis* of the required simulation and use the *SysML4HLA* profile to annotate the PIM in order to enrich such a model with the information required to derive the HLA-based simulation model. Specifically, the *SysML4HLA* profile allows one to specify both how the system has to be partitioned in terms of Federation/Federates and how system model elements have to be mapped to HLA model elements such as *object classe*s and *interaction classes.*

Then, the *design simulation environment* step is executed. This step takes as input the marked PIM and executes the *SysML-to-HLA model-to-model* transformation in order to automatically obtain a UML model that represents the HLA simulation model. Such a model is annotated with the stereotypes provided by the *HLA* profile and, according to the MDA terminology, is referred to as the *platform specific model* (*PSM*). The design simulation environment step is also concerned with the discovery of existing federates to be integrated in the distributed simulation.

The *develop and integrate simulation environment* step is then executed to implement the distributed simulation. The simulation code, specified by the use of the *HLA Development Kit software Framework (DKF)*, is generated through the execution of the *HLA-to-Code model-to-text* transformation. This step also includes the coding activities needed to integrate the existing Federates identified in the previous step.

Finally, the distributed simulation is executed and the results are evaluated to check whether or not the predicted system behavior satisfies the user requirements and constraints. In the positive case, the validated SysML-based system specification can be used to drive the possible design and implementation of the system. Alternatively, the system specification has to be revised.

The *HLA Development Kit software Framework (DKF)* is provided by the University of Calabria in cooperation with the NASA JSC (Johnson Space Center) along with related documentation, user guide and reference examples [8]. The *DKF* is implemented in the Java language and is based on the following three principles: (i) *Interoperability, DKF* is fully compliant with the IEEE 1516-2010 specifications; as a consequence, it is platform-independent and can interoperate with different HLA RTI implementations (e.g. PITCH, VT/MÄK and CERTI [9]); (ii) *Portability and Uniformity*, *DKF* provides a homogeneous set of APIs that are independent from the underlying HLA RTI and Java version. In this way, developers could decide the HLA RTI and the Java run-time environment at development-time; and (iii) *Usability*, the complexity of the features provided by the *DKF* framework are hidden behind an intuitive set of APIs.

A side advantage of the proposed method is that the same approach can be adopted to eventually generate both the operational system and the distributed simulation system starting from the same model specification.

The next section introduces the reference example that is used hereinafter to illustrate the details of the method steps. It shows the steps that go from the initial system specification down to the development of the distributed simulation code.

3 Reference Example

The simulation scenario deals with a *docking system* and concerns a situation in which an Automated Transfer Vehicle (ATV) is approaching the International Space Station (ISS) to dock on it. Although the docking procedure is handled by the ATV, to allow the ATV to dock safely, both the ISS and the Mission Control Center on the earth have to be informed in order to monitor and face with possible critical situations. When the ATV begins the docking operations, it sends an "Approach—started" message. The personnel on the ISS and the Mission Control Center on earth react to the situation accordingly. During the approaching phase, the ATV constantly monitors its trajectory and distance from the assigned docking point on the ISS and acts accordingly to reach the target; moreover, the ATV constantly sends information about its position, acceleration and velocity. When the ATV docks on the ISS it sends a "Docking—completed" message.

The following subsection applies the MONADS method illustrated in Sect. 2 to the considered docking system.

4 Method Application

This section describes the various steps needed to carry out the proposed model-driven method and thus generate the distributed simulation code of the docking system.

The following subsections describe the different steps in more detail, according to the method overview illustrated in Sect. 2.

4.1 SysML Modeling

As aforementioned, the proposed method is carried out through several steps, the first of which includes the definition of the system model by the use of the SysML notation.

For the purposes of this discussion, the study is limited to those diagrams that are necessary to obtain the simulation model and the code of the distributed simulation.

Specifically, the SysML model of the docking system is composed of the following diagrams: (i) *a block definition diagram (BDD)*, which specifies the structural view of the system. The diagram shows the docking system components (e.g., ATV, ISS and their internal elements) and their structural relationships; (ii) *a set of sequence diagrams (SDs)*, which specify the behavioral view of the system. Such diagrams depict the ordered set of interactions between different system components.

4.2 *From SysML to HLA-Based UML*

At the second step, the stereotypes of the *SysML4HLA* profile are used to annotate the aforementioned SysML model, so as to drive its mapping to the HLA-based UML model. Such stereotypes add information needed for the automatic mapping between SysML domain elements and the corresponding HLA domain elements. Figures 2 and 3 show the BDD for the considered docking system and a SD specifying the interactions between the AVT Federate component and the AVT Object Class component, respectively.

The resulting marked PIM is taken as input by the automated *SysML-to-HLA* model-to-model transformation, which yields as output the UML model of the corresponding HLA-based distributed simulation. The resulting UML model is composed of the following diagrams: (i) *a set of sequence diagrams*, which specify the behavioral view of the HLA simulation model; (ii) *a component diagram,* which describes the structural view of the model; (iii) *a component diagram,* which shows the publish/subscribe associations between *federates*, *ObjectClass* and *InteractionClass* HLA elements (see Sect. 1).

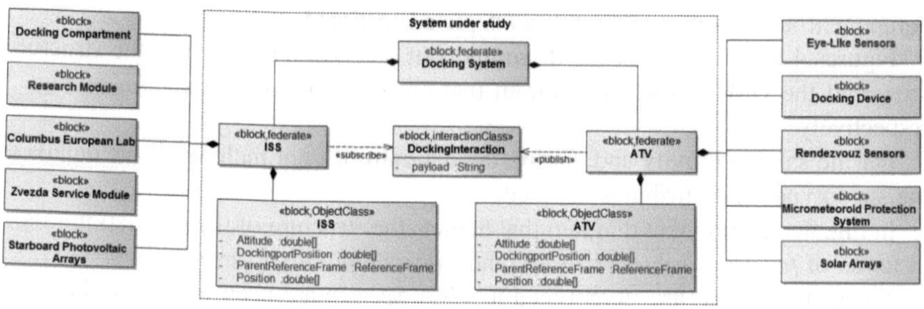

Fig. 2 SysML BDD of the docking system

Fig. 3 Interactions between ATVFederate and ATVObjectClass

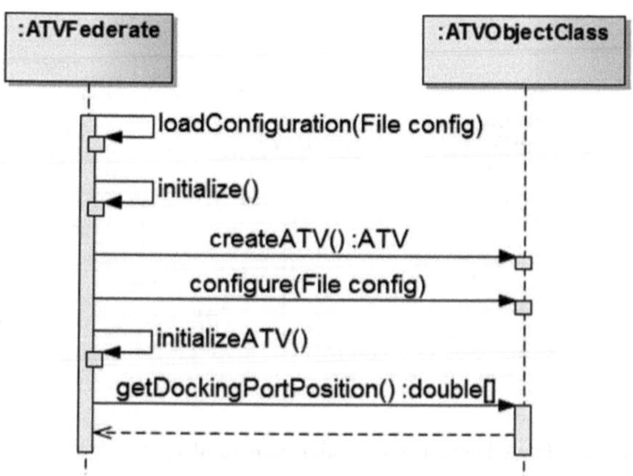

As stated in Sect. 2, the *SysML-to-HLA* model-to-model transformation has been specified by the use of the QVT/Operational Mappings (QVT-O) language [6], the standard language for defining operational transformations consisting of a set of mapping functions, which specify the mapping rules by the use of conventional imperative primitives. Among the *mapping rules* specified to generate the structural model, each block element in the SysML source model is mapped to a *federate* element, an *ObjectClass* element or an *InteractionClass* element in the target model, depending on the relevant SysML4HLA stereotype (i.e., the role played in the source model). As an example, the ATV block element stereotyped as <<*federate*>> in the SysML BDD of Fig. 2 is mapped to the UML ATV component stereotyped as <<*federate*>> in the UML component diagram of Fig. 4.

Similarly, for the behavioral model, a set of mapping rules enable the transformation of the SysML sequence diagram in the source model into a UML sequence diagram in the target model. More precisely, the interaction between a couple of blocks stereotyped as ≪*federate*≫ and ≪*objectclass*≫ in the source model is mapped to an interaction between the UML component representing the federate block and the DKF component that wraps the RTI. Such an interaction consists of a set of predefined messages stereotyped as ≪*initialization*≫, ≪*message*≫ or ≪*action*≫, to represent the behavior of a federate interacting with its component and/or other federates, as depicted in Fig. 5.

Figures 4 and 5 show the UML component diagram that specifies the structural view and the UML sequence diagram that corresponds to the diagram of Fig. 3, respectively.

For the sake of brevity and clarity, the diagrams do not include all the details of the stereotypes and attributes adopted.

In Fig. 4, a component diagram that depicts the structure of the simulation model in terms of *federates*, *object classes*, and *interaction classes* can be observed.

The elements that are stereotyped as *HLA Service* are used to enable the interaction between federates, conventionally carried out by the use of the HLA RTI (runtime infrastructure) component. In this paper case, the RTI is replaced by the wrapper components provided by the HLA Development Kit.@@

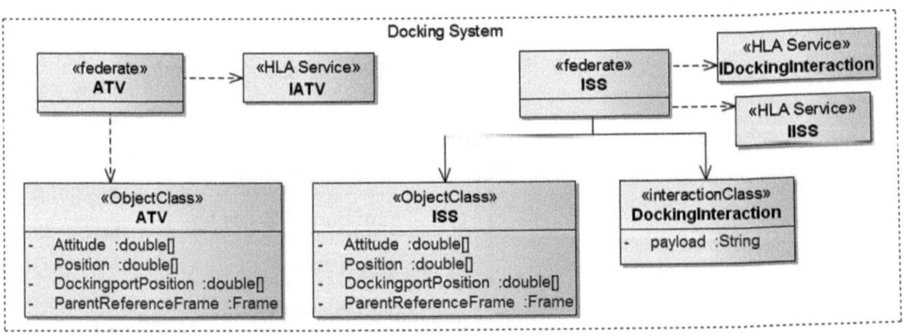

Fig. 4 HLA-based UML model: structural view

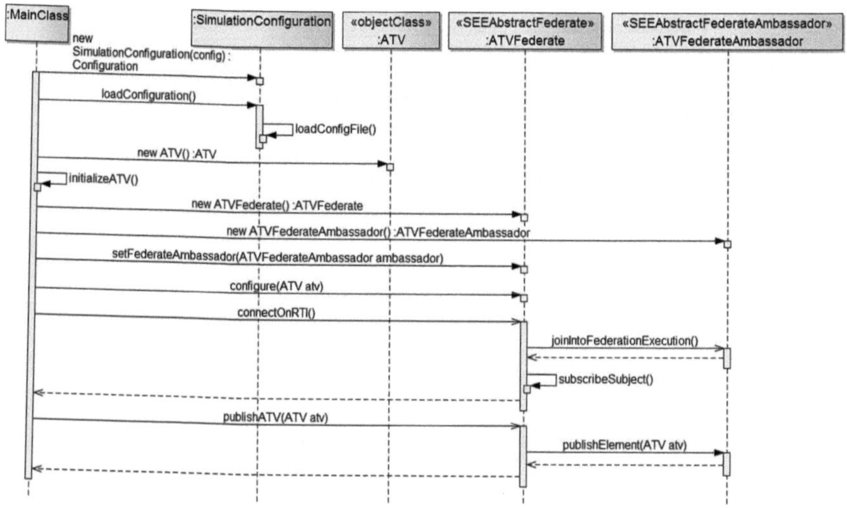

Fig. 5 HLA-based UML model: interaction view

4.3 *From HLA-Based UML to DKF-Based Code*

At the last step, the HLA-based UML model produced in the previous step is taken as input by the automated *HLA-to-Code* model-to-text transformation, which yields as output a considerable part of the distributed simulation code. As stated in Sect. 2, the *HLA-to-Code* transformation has been specified in the MOFM2T language, which adopts a template-based approach, wherein a template specifies a code template with placeholders for data to be extracted from models.

Specifically, the *HLA-to-code* transformation allows one to generate a template of the Java classes that contains the class structure, including constructors, methods and attributes, declarations and exception management, and most of the required HLA-related code, such as data type definitions and RTI interaction methods. The only code that has to be manually added is the one implementing the federate simulation logic. Specifically, the DKF allows developers to focus on the specific behavioral aspects (both the proactive and reactive simulation logic, see Listing 1) of their own HLA Federates rather than dealing with the common HLA functionalities, which are managed by the DKF core components. In addition, the behavior of a Federate follows a well-defined life cycle provided and managed by the DKF [10] thus producing more reliable and easy to maintain simulation code.

```java
public class ATVFederate extends SEEAbstractFederate implements Observer {

  private ATV atv = null;

  public ATVFederate(SEEAbstractFederateAmbassador seefedamb, ATV atv) {
     super(seefedamb);
     this.atv  = atv;
  }

  public void configureAndStart(Configuration config) {
     // 1. configure the DKF framework
     super.configure(config);

     // 2. Connect on RTI
     super.connectOnRTI("crcHost="+config.getCrcHost()+"\ncrcPort="+config.getCrcPort());

     // 3. The Federate joins the Federation execution
     super.joinIntoFederationExecution();

     // 4. Subscribe the Subject
     super.subscribeSubject(this);

     // 5. publish the ATV object on RTI
     super.publishElement(atv);
     super.subscribeReferenceFrame(FrameType.EarthCentricInertial);
     super.subscribeInteraction(DockingInteraction.class);

     // 6. Execution-loop
     super.startExecution();

     System.out.println("Press any key to disconnect the federate from the federation");
     new Scanner(System.in).next();
     diconnectFromRTI();
  }

  protected void doAction() {
     // Proactive Simulation logic.
  }

  public void update(Observable arg0, Object arg1) {
     // Reactive Simulation logic.
  }
}
```

Listing. 1 The DKF-based code of the ATVFederate

5 Related Work

This section reviews the existing literature dealing with both the use of SysML in the Modeling & Simulation (M&S) domain and the modeling/development of HLA-based distributed simulation systems.

As regards the use of SysML in the M&S context, a significant contribution that specifically addresses the generation of Java/HLA code from SysML specifications can be found in [11]. This work extends and improves such a contribution both on the method side, which is now designed according to the DSEEP, and on the model transformation side, which now exploits the advantages of using the *HLA Development Kit software Framework* (*DKF*) rather than a conventional HLA implementation.

More generally, several contributions are available that propose the use of SysML as a notation suitable not only for defining systems specification but also for supporting system simulation activities, such as [12] and [13] in which SysML is used as a notation to support the simulation-based design of systems, in order to derive executable parametric models and simulation-specific languages, respectively.

Differently from the aforementioned contributions, this paper describes a model-driven method to generate an HLA-based implementation of a distributed simulation software, starting from a SysML specification.

As regards the issue of supporting the implementation of simulation systems, contributions that apply a model-driven paradigm in the M&S domain can be found in [14] and [15], which propose a method to generate a Java/HLA-based implementation of a distributed simulation software from a UML system model and the main theoretical concepts behind the application of MDA to HLA, respectively.

Differently, this paper illustrates the design and implementation of a model driven method to reduce the gap between the SysML-based system specification and the HLA-based distributed system implementation.

As regards the modeling/development of HLA-based distributed simulation systems, several commercial and research efforts aim at providing integrated toolchains for creating and simulating complex systems by using specialized modeling tools and methodologies. For MATLAB/Simulink different packages and toolboxes are available for implementing HLA simulators such as the Forwardsim HLA Toolbox for MATLAB [16] and the HLA/DIS Toolbox for MATLAB and Simulink [9].

Another tool that enables developers to effectively manage the structure and assets of a HLA Federate starting from a FOM (Federation Object Model) file is the PITCH Developer Studio [9]. A domain-specific HLA software framework was created by the Danish Maritime Institute (DMI) [17] to provide mechanisms that simplify the development of real-time simulators. Other HLA frameworks are based on GRID-computing infrastructure [18].

The *HLA Development Kit* and its software framework (*DKF*), used in the proposed method for generating the HLA-based simulation code (see Sect. 2), differ from the above mentioned solutions in several aspects. In particular, differently from a proprietary and commercial solution that requires tool-specific knowledge and training, the *HLA Development Kit* is an open source project released under the open source LGPL license and can be freely and easily customized and/or extended to cover and deal with both domain independent and domain-specific aspects. In addition, the DKF provides advanced facilities that allow keeping the code compact, readable and reliable. As an example, Java annotations are used to directly inject the structure of a HLA Federate in the Java code. These metadata are used by the core components of the DKF at run-time to inspect and check HLA objects according to its definition in the FOM. The above-sketched capabilities showed a great benefit not only for expert HLA developers but also for HLA novice practitioners as were the undergraduate students involved in the Simulation Exploration Experience (SEE) project led by NASA and which involves several U.S. and European Institutions [10, 19].

6 Conclusions

Modern large-scale systems or systems of systems require the adoption of distributed simulation approaches to properly take into account the inherent complexity of such systems.

This paper has introduced an innovative and automated method (denoted as MONADS) that makes easier for systems engineers the use of distributed simulation techniques, without asking them to explicitly deal with the intricacies and difficulties of currently available standards and technologies (HLA in this paper case).

The contribution of the paper is twofold. On the one hand, it introduces a model-driven method based on the execution of model transformations that automatically map the abstract representation of a system, specified in SysML, into an intermediate HLA-based software model, specified in UML, down to the final code of the HLA-based distributed simulation. On the other hand, it exploits an innovative software framework, the HLA Development Kit software Framework (DKF), that allows one to appropriately handle common HLA issues thus making it easier to get to the final code of the distributed simulation.

The proposed approach allows one to automatically obtain a significant portion of the final HLA-based code, by limiting the manual activity to the implementation of the federate simulation logic, and can be effectively used even by systems engineers who are not familiar with the HLA standard.

Work is in progress to evaluate the opportunity of incorporating additional transformations that would allow increasing the portion of the automatically generated code at the expenses of an extra modeling effort.

References

1. Fujimoto, R.M.: Parallel and Distributed Simulation Systems. Wiley (2010)
2. IEEE Standard for Modeling and Simulation (M&S) High Level Architecture (HLA)—Federate Interface Specification, IEEE Standard 1516-2010
3. Kuhl, F., Weatherly, R., Dahmann, J.: Creating Computer Simulation Systems: An Introduction to the High Level Architecture. Prentice Hall (1999)
4. Atkinson, C., Kuhne, T.: Model-driven development: a metamodeling foundation. IEEE Softw. **20**(5), 36–41 (2003)
5. OMG. MDA Guide, version 1.0.1 (2003)
6. OMG. Meta Object Facility (MOF) 2.0, MOF Query/View/Transformation 1.0, MOF Model to Text Transformation Language 1.0 (2008)
7. IEEE Recommended Practice for Distributed Simulation Engineering and Execution Process (DSEEP). IEEE Standard 1730-2010 (2011)
8. HLA Starter and Development Kit, https://code.google.com/p/see-hla-starterkit/ (2015)
9. HLA RTI implementations: Pitch portable RTI, http://www.pitch.se; MÄK, VR-Forces, http://www.mak.com/; CERTI, http://savannah.nongnu.org/
10. Anagnostou, A., Chaudhry, N.R., Falcone, A., Garro, A., Salah, O., Taylor, S.J.E.: A prototype HLA development kit: results from the 2015 simulation exploration experience. In: Proceedings of the SIGSIM PADS 2015, London, UK, June, 10–12, 2015

11. Bocciarelli, P., D'Ambrogio, A., Fabiani, G.: A model-driven approach to build HLA-based distributed simulations from SysML models. In: Proceedings of SIMULTECH 2012, pp. 49-60. Rome, Italy, July, 28–31, 2012
12. Peak, R.S., Burkhart, R.M., Friedenthal, S.A., Wilson, M.W., Bajaj, M., Kim1, I.: Simulation-based design using SysML—part 1: a parametrics primer. In: Proceedings of the INCOSE International Symposium, vol. 17, no. 1, pp. 1516–1535 (2007)
13. Paredis, C.J.J., Johnson, T.: Using OMG's SysML to support simulation. In Proceedings of the Winter Simulation Conference (WSC '08), pp. 2350–2352 (2008)
14. D'Ambrogio, A., Iazeolla, G., Pieroni, A., Gianni, D.: A model transformation approach for the development of HLA-based distributed simulation systems. In: Proceedings of SIMULTECH 2011, pp. 155–160. Noordwijkerhout, Netherlands, July, 29–31, 2011
15. Haouzi, H.E.: Models simulation and interoperability using MDA and HLA. In: Proceedings of the IFAC/IFIP International conference on Interoperability for Enterprise Applications and Software (2006)
16. The Forwardsim HLA Toolbox for MATLAB, http://www.forwardsim.com/ (2015)
17. Villimann, O.: CTO Project. HLA Framework, Danish Maritime Institute (1999)
18. Xie, Y., Teo, Y.M., Cai, W., Turner, S.J.: Towards grid-wide modeling and simulation. (2005)
19. Simulation Exploration Experience (SEE) project, http://www.exploresim.com/ (2015)

Maintenance as a Cornerstone for the Application of Regeneration Paradigm in Systems Lifecycle

Laëtitia Diez, Pascale Marangé, Frédérique Mayer and Eric Levrat

Abstract The circular economy is an economy, firstly, considering the natural resources as finite and the non-existence of waste, secondly, assimilating the industrial system as a natural system and, finally, emphasizing the paradigm of regeneration. Nevertheless, this paradigm is not clearly defined and this paper aims to found it by proposing solutions to its implementation in the industrial world. The proposal is based on a comparison between the natural system and the industrial system by using the trophic organization model and their elements. Then, the maintenance process is seen as a key element of regeneration. Finally, the notion of nutrient is studied and taken into account in an industrial process.

Keywords Sustainable development · Circular economy · Regeneration · System lifecycle · Maintenance

1 Introduction

Further to the analysis on the climate change, the greenhouse gases, the exhaustion of the natural resources and the increase of waste…, the European political powers have implemented a strategy titled "Europe 2020" [1]. Just like for the United

---

L. Diez (✉) · P. Marangé · E. Levrat
CNRS CRAN UMR 7039, Université de Lorraine, BP 70239 boulevard des aiguillettes, F-54506 Vandoeuvre, France
e-mail: laetitia.diez@univ-lorraine.fr

P. Marangé
e-mail: pascale.marange@univ-lorraine.fr

E. Levrat
e-mail: eric.levrat@univ-lorraine.fr

F. Mayer
ENSGSI, ERPI, Université de Lorraine, EA no 3767 8 Rue Bastien Lepage, 54010 Nancy, France
e-mail: frederique.mayer@univ-lorraine.fr

G. Auvray et al. (eds.), *Complex Systems Design & Management*,
DOI 10.1007/978-3-319-26109-6_14

Nations Conference on Sustainable Development (UNCSD) or Rio +20, this strategy is based on the three pillars (economy, environment, social) of sustainable development. This development is defined as a "development that meets the needs of the present without compromising the ability of future generations to meet their own needs" [2] and aims to change the economy.

Faced with this development, the linear model (take-make-dump) becomes obsolete and gives way to circular model (make-use-return infinitely). This new economy is increasing and is promoted by diverse organization, such as the "Ellen MacArthur Foundation" in United Kingdom, the "Institut de l'Economie Circulaire" in France, the international certification "Cradle to Cradle®", etc.

The circular economy considers that the natural resources exist in finite quantities and that the waste do not exist [3]. Thus, the general idea is to create closed loops by reusing constantly the manufactured goods, based on the system thinking, the use of renewable resources and the fact that waste equal resources. This approach forces us to change the paradigm and to move towards the "regeneration" paradigm. This one relies on several schools of thought, like:

- Biomimicry [4] is the study of nature to design goods and systems in a sustainable way. This concept has promoted products. For example, the high-speed train is inspired by the form of some birds, the glue by the adhesive capacity of mussels, the swimsuit by the skin of sharks, the new energy production by the photosynthesis, etc.;
- Industrial Ecology [5] is the study of material and energy flows through the industrial system to add value to waste from one firm as resources for one or more other firms. This concept rely on the optimization calculations and on the Life Cycle Analysis. Industrial Ecology considers the industrial system as an ecosystem to create a closed-loop operating as the nature [6]. The Kalundborg area [7] is a good example of industrial symbiosis, which aims to optimize the water use, to save energy and to reuse waste;
- Regenerative Design is the study of environment and community of a place to design in harmony with them [8]. This design involves seeing the environment, the community, and the systems as a whole [9]. This design goes further than the sustainable design in regenerating the end-of-life systems. A product is regenerative if and only if it is 100 % recycled and recyclable and improves the environmental and human conditions;
- Cradle to Cradle [10] is the popular term of regenerative design and uses a biomimicry approach to design goods. This concept considers the materials as nutrients (organic and technical). The biological nutrients are organic materials that can return healthily in the natural cycle. The technical nutrients are non-toxic, non-harmful synthetic materials for the natural environment, but they cannot return in the biosphere and must go back in the technical cycle.

With all this concepts, the regeneration paradigm is difficult to set up. Indeed, in the best-case scenario, every firm would be able to regenerate its own goods and waste to produce new secondary matters and energy (Fig. 1a). This is workable by the future companies, but for the existing firms, this is more difficult. Creating firms

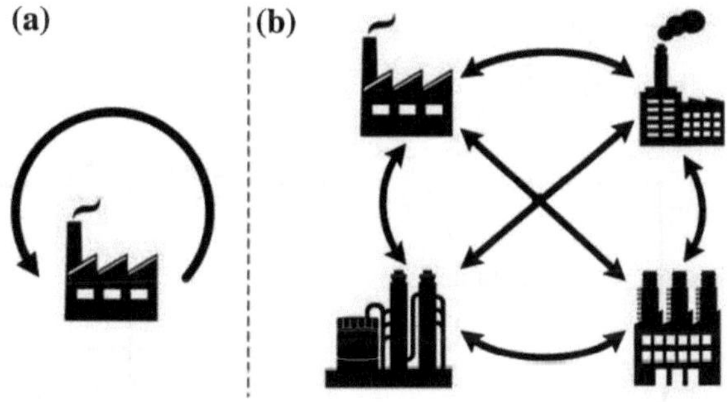

Fig. 1 Auto-regeneration (**a**) and regeneration by network of companies (**b**)

networks is an interesting idea. Each of these firms is able to regenerate the goods of one or several companies and to consume the energy produces by itself or other companies (Fig. 1b).

The purpose of this paper is to determine if the regeneration paradigm is relevant to the industrial world and to propose a way to develop it in the industry, based on the maintenance process.

For this, the Sect. 2 describes, first, what is the regeneration in nature through the food web and, secondly, transposes the trophic organization to the industrial world. Then Sect. 3 identifies the processes, which allow regenerating a nutrient in the product lifecycle. The Sect. 4 defines the elements necessary to nutrient regeneration. A discussion/conclusion is proposed in Sect. 5.

2 Regeneration Concept

Currently, biosphere and technosphere (part of biosphere affected by modifications of the human origin) are seen completely independently of one another. However, in the reality and from the point of view of the circular economy, the biosphere contains the technosphere, which affects constantly this first. In fact, the technical cycle of the technosphere affects the natural cycle, which evolves in the biosphere. To understand these influences, the different elements allowing the interaction of cycles and their regeneration must be identified.

Biology defines the verb "regenerate" as a process "to grow again". In other words, the nature regenerates continuously and does not know the notion of waste, contrary to the industry. The principle of biomimicry is a good starting point to understand the regeneration paradigm. Especially as, in the biosphere, the exchanges are circular and as the industry uses resources from nature.

2.1 Natural Regeneration

The food web is an excellent example of circularity and regeneration. Indeed, Ecology emphasizes that the trophic organization allows at each population of

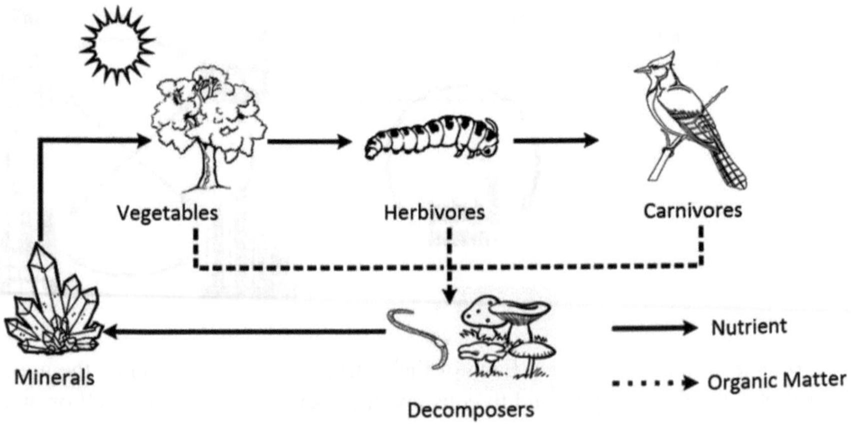

Fig. 2 Trophic organization (natural)

living being to replace each other naturally. This organization is based on the exchange of nutrient organic, mineral and energizing between the trophic levels, i.e., that each living organism feeds it another [11]. A nutrient is a nutritive substance that sustains the life. Actually, vegetables (primary producers) feed of solar energy and mineral nutrients. Herbivores (primary consumers) eat vegetables and the carnivores (secondary consumers) eat herbivores and other carnivores. Each of these categories generates organic matters, which are eaten by decomposers (ex: bacteria, fungi, worms). These decomposers produce mineral nutrients to feed vegetables, and so on and so forth (Fig. 2). So all resources regenerate except the solar energy that is a renewable energy for a long time, at the human scale.

In this organization, the fundamental elements are decomposers. In fact, these organisms transform an unusable matter (organic matter) by producers and consumers in another matter (nutrient) usable by vegetables. Without them, the trophic organization would not be circular but linear. That would result by an accumulation of organic matters, an exhaustion of vegetables and by the extinction of species.

2.2 *Industrial Regeneration*

Contrary to the natural cycle, the technical cycle is linear. In fact, the firms of primary sector business exploit the resources of nature (organic and mineral) to supply the firms making goods (secondary sector). This sector supplies tertiary sector. Each of those sectors produces waste. Like so, the industrial world accumulates waste and exploits the finite resource of nature, and that is the problem. Indeed, the storage of waste costs much and the natural resources make rare. To solve this problem, the nature must inspire industry. Especially as the nature does not need storage and is able to regenerate its resources.

If one compares the behavior of the natural system with the industrial system, we note that there is no business sector, which treats waste to transform them in the raw material. In the nature, the decomposers play this job. To become regenerative and circular, the industry must organize decomposers.

Thus, the parallel with the nature allows assimilating the primary sector business with the primary producers, the secondary sector with the primary consumers and the humans activities with secondary/tertiary consumers [12]. Decomposers allow regenerating waste from sectors and transforming them in raw secondary matters for each sector or in natural resources for nature and the primary sector (Fig. 3).

Moreover, as the biosphere, several living organisms compose the technosphere. A company this is an (artificial) complex system linked and organized. This organism must consume resources, i.e. nutrients, to live, develop and prosper. In the industrial world, resources correspond to materials, energy, services, staff and knowledge. The secondary sector produces objects that are resources, i.e., nutrients for human activities. An object is often a set of objects. Consequently, a nutrient is a set of nutrients.

Finally, as the trophic organization, business sectors trade nutrients and produce waste regenerated by decomposers. However, these sectors consume different types of nutrients from the biosphere (organic and mineral) and technosphere (technical). Wood and leather are organic nutrients, water and iron are mineral nutrients, and paper and PVC are technical nutrients.

The natural decomposers can be used to regenerate organic nutrients and to create mineral nutrients for vegetables. Industry must organize technical decomposers to regenerate technical nutrients and to create all natures of nutrients (organic, mineral and technical). The accumulation of mineral nutrients, from technical decomposers, are artificial mines for the primary sector (Fig. 4). For this figure, we consider a perfect cycle where all waste are regenerated.

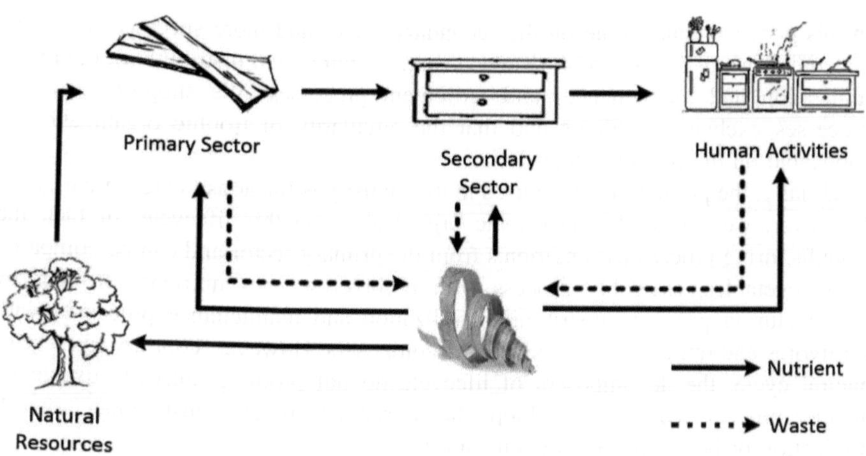

Fig. 3 Trophic organization (industrial)

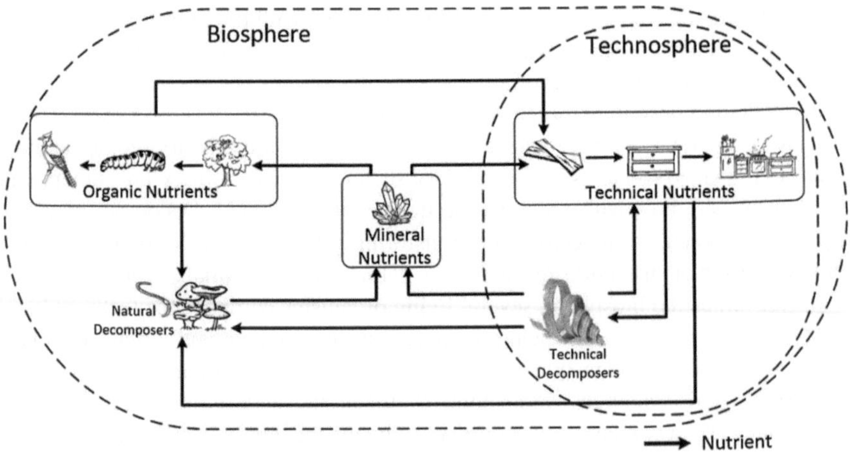

Fig. 4 Interaction between natural and technical cycle

Biosphere and technosphere represent and form a set of ecosystem mutually dependent and accommodate living organisms. In fact, the notion of nutrient allows making a link between the different elements of each cycle and between the cycles themselves. By adding the technical decomposers, in technosphere, the goods become regenerable and allow reducing waste and the exploitation of natural resources. Moreover, the use of natural decomposers to regenerate waste of technosphere allows giving back nutrient to exploit.

3 Product Lifecycle

In this part, we concentrate on the secondary sector and more specifically, on the material flows of production (manufacturing, assembly and distribution), operation (utilization and maintenance) and retirement processes. We suppose also that processes exchange nutrients and that the circularity of trophic organization is applicable on the product lifecycle.

Usually, the product lifecycle used in the industry is linear as the technical cycle. Moreover, we can apply the trophic organization on this lifecycle. In fact, the manufacturing process uses nutrients from the primary sector and can be compared with vegetables, assembly process with herbivores (uses nutrients product by manufacturing process), distribution, utilization and maintenance processes with carnivores and retirement process with decomposers. However, contrary to these of natural cycle, the decomposers of lifecycle do not produce nutrients usable by another process. To close the loop, the retirement process must return towards production or better towards the customer.

3.1 Closed-Loops

References [3, 13] suggest creating a loop between retirement and all other processes. In other words, the retirement process (technical decomposers) is able to reuse, remanufacture and recycle a nutrient. The maintenance process is also a decomposer, but only replacements, adjustments and repairs actions. Furthermore, in the general system theory [14], a process can be defined in a "time-space-shape" (TSS) frame reference. Nevertheless, this referential is not enough to define a decomposer. In fact, decomposers can alter the nutrient functionality. If we compare a nutrient going in and out process, this nutrient is always affected in the time, may be transported and/or modified on its shape, and may keep its functionality. Moreover, each object by circulating in a process is characterized by "state variable" relating to TSS referential, i.e., temporal, spatial and shape properties [15]. In the same way, as for the flows, a nutrient has these properties, which will develop in the next section. Therefore, technical decomposers can be defined as (Fig. 5):

- Replacement/Repair process: transformation of time, and preservation of nutrient functionality;
- Reuse process: transformations of time and space, and preservation of nutrient functionality;
- Remanufacture process: transformations of time, space and shape, and non-preservation of nutrient functionality. But the nutrient parts keep their functionality;
- Recycling process: transformations of time, space and shape, and non-preservation of nutrient functionality. The nutrient parts do not keep their functionality.

In this list, only technical decomposers are cited to regenerate technical nutrients, but the industry uses also organic nutrients. Natural decomposers must be added at this list to regenerate those nutrients and make transformations of time,

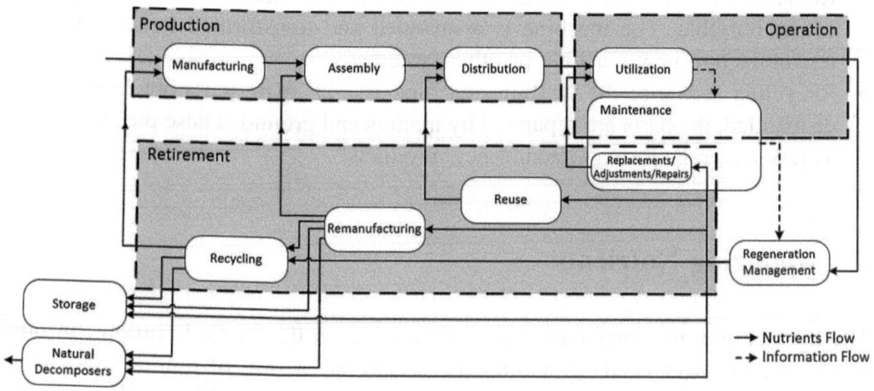

Fig. 5 Closed-loops of product lifecycle

space, and shape on organic nutrients, without preservation of nutrient functionality. The vegetables can use these regenerated nutrients. However, with current technologies, a decomposer can regenerate not all nutrients; this nutrient will be stored and kept in the technosphere.

After the identification of decomposers, we must find how to select an appropriate decomposer to regenerate nutrient and determine when regenerating it. A regeneration manager is necessary, and to make this job, we propose to add a regeneration management process (Fig. 5). This process rests on the information provided by the maintenance process. In fact, the maintenance process is a process which aims to fix or restore an item in a state in which it can perform its function with dependability [16]. Therefore, it has information on the system state of health. Moreover, in the product lifecycle, this process is situated between production and retirement phases, i.e., before decomposers, what makes it essential to support the nutrients regeneration management.

3.2 *Example: Washing Machine*

To illustrate the different cycles and decomposers, we are interested in a simple example: a washing machine used by a laundry.

- Replacement/Repair scenario: the washing machine is faulty but can be repaired and uses again by the firm;
- Reuse scenario: the washing machine uses too many resources (water, electricity, detergent) or does not adapt to new requirements of the laundry (cold wash and not hot wash), the quality of washing is deteriorated, etc., in other words, loss of technical, energy and economic performances. Hotels, hospital or old people's home, which need the only functionality of the machine and not its performances can reused this machine. A laundry can become also bankrupt and put its machines on the second-hand market.
- Remanufacturing scenario: the washing machine is faulty (cannot be repaired) or not reusable. The machine is dismantled and transformed in spare parts to remanufacture new machine or other systems;
- Recycling scenario: the washing machine and its spare parts are faulty. After dismantled, the parts are separated by matters and ground. These pieces are used as raw materials to manufacture new products.

4 Managing Nutrients

The regeneration manager must have information on the studied nutrient, in other words, the manager needs properties to identify the nature of nutrient and which decomposers used according to different properties of sustainable development and the market needs (nutrients needed by customers with a good added value). The nutrient has properties from processes so-called technical. However, the nutrient definition focuses mainly on the environmental impacts of the nutrient. This requires introducing ecological properties.

The NIAM/ORM language [17], with NORMA plug-into Microsoft® Visual Studio® 2010 Ultimate, is used to model the link between spheres and nutrient properties. This language allows writing out elementary facts.

4.1 Definitions

A nutrient is composed of two types: biological nutrients and technical nutrients. Previously, we have seen that the biological nutrients are organic materials that can return healthy in the natural cycle. However, this definition considers only organic materials while industry uses mineral materials too. To take account of these materials, we will talk about natural nutrients instead of biological nutrients. Regarding the technical nutrients, they are non-toxic, non-harmful synthetic materials for the natural environment, but they cannot return in the biosphere and must stay in the technical cycle. This definition is too restrictive. In this paper, the technical nutrient will be a nutrient non-natural that harms the biosphere that the technical cycle must use as long as possible.

4.2 Elements to Identify Types of Nutrient

Two stages are necessary to identify the type of nutrient. In first, we must determine if the nutrient harms or not the biosphere and, if the natural decomposers can regenerate it. For this, environmental properties are linked to the biosphere and ecological properties to nutrient. The two categories of properties are identical. A fixed threshold allows comparing the values of an ecological property (carbon footprint, aquatic toxicity, etc.) of nutrient with the environmental properties (Fig. 6). In the Cradle to Cradle® certification, a list of environmental health endpoints [18] is available with threshold varying as the function of certification level desired. This list is based on harmful chemical elements for the environment and human and on their quantities. This certification provides a fulcrum to identify the non-harm of nutrient for the biosphere.

In second, we must determine if a nutrient can be regenerated or not by the technical decomposers. For this, the approach is the same for ecological properties but with technical properties (dimensions, weight, global cost, etc.). Their values are compared with the thresholds of regeneration properties (of technosphere) (Fig. 7).

In other words, the properties lied to spheres define a nutrient. These properties are compared with the properties of each cycle. However, contrary to thresholds, the values of the nutrient change in the time and all through processes. In fact, the

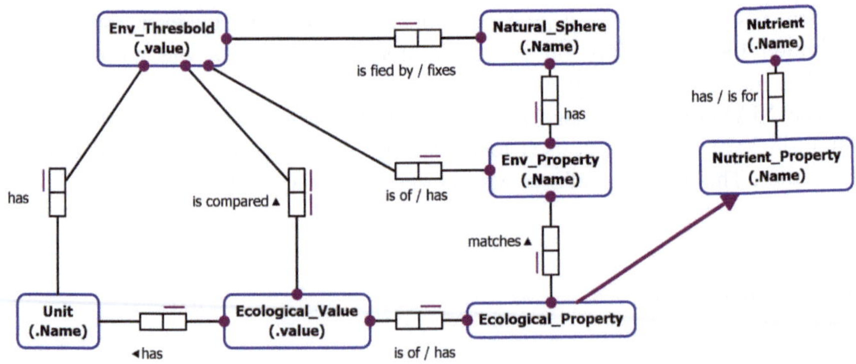

Fig. 6 Modeling of the environmental and ecological properties

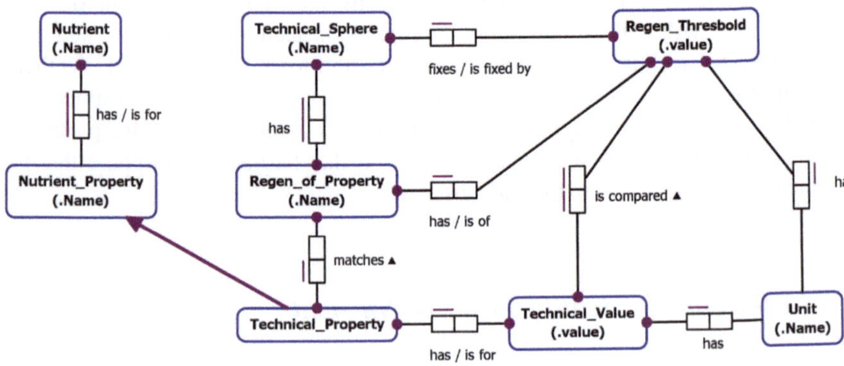

Fig. 7 Modeling of the regeneration and technical properties

nutrient evolves in the TSS referential and the processes transform it. These affect the properties of nutrient and their values.

After this modeling, we make the distinction between technical and natural nutrient, and between regenerable and non-regenerable nutrient.

4.3 *Types of Nutrient*

Taking into consideration that the nutrient properties can be compared with those of biosphere and technosphere, we have created a generic pattern (Fig. 8) to realize this comparison. Thus, if a nutrient of type N satisfies all properties of given sphere, then the nutrient is type N1, return to the sphere and a decomposer (natural for biosphere and technical for technosphere can regenerated by decomposers) it, else that is a nutrient of type N2. The generic pattern is applied on the nutrient N2 that

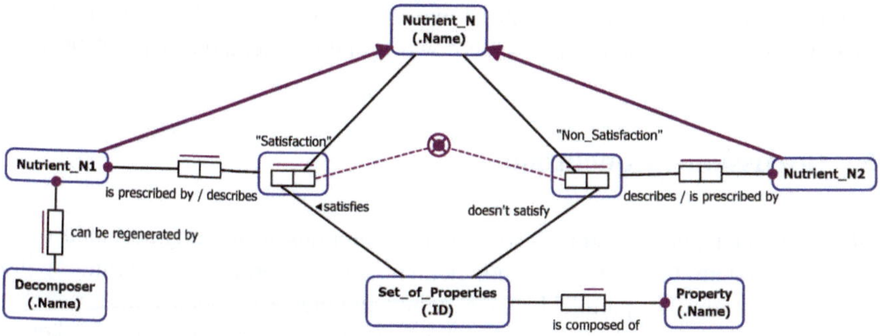

Fig. 8 Generic pattern to identify the type of nutrient

becomes nutrient N. This pattern is executed until the obtaining a non-regenerable nutrient by either sphere.

A possible instantiation is to compare nutrient ecological properties with environmental properties. If a nutrient does not satisfy at least one environmental property, then it is a pollutant for the environment and defined as technical, else it is natural and regenerable by natural decomposers. Likewise, if a nutrient respects all regeneration properties, then it is regenerable by the technical decomposers and remains in the technical cycle, else it is a non-regenerable nutrient.

Contrary to Refs. [3, 10], we consider the existence of "waste" to take account of existing goods and not only on those to design. However, these waste are minimized and must be regenerated a maximum. A non-regenerable nutrient by the technosphere is waste and oblige us to think about its return in the natural cycle.

A non-regenerable nutrient identified as natural is healthy for the biosphere and can return there. Nevertheless, if the nutrient is technical, then it is harmful. Therefore, a harmful nutrient must necessarily to get good properties back, in other words, be confined until technologies enable correctly regenerated it. A cleansed nutrient will able to return in one of the cycles.

This typology allows knowing the nutrient origin and the cycle able to regenerate the studied nutrient. All types of nutrients are defined in the following list:

- Nutrient: element used in an industrial process;
- Technical nutrient: nutrient harming to the biosphere;
- Natural nutrient: nutrient non-harming to the biosphere that can return there;
- Biological nutrient: natural nutrient of organic origin;
- Mineral nutrient: natural nutrient of mineral origin;
- Regenerable nutrient: nutrient satisfying all technical properties and can return to the technical cycle;
- Non-regenerable nutrient: nutrient not satisfying at least one technical properties;
- Healthy nutrient: natural and regenerable nutrient;
- Harmful nutrient: technical and non-regenerable nutrient.

Of course, a nutrient that is natural and technical-regenerable can return in the two cycles, and this choice will come down to the regeneration manager process.

5 Discussion—Conclusion

Observing the natural system allows us to find solutions at currently problems of the industrial system, to identify missing elements to transform a linear economy in a circular economy. The parallel between nature and industry evidence the absence of decomposers in the technical cycle. However, the decomposers are the key of regeneration in the natural cycle. Indeed, they transform the organic matters (living organism waste) in mineral salts, elements used by vegetables to grow. Therefore, to regenerate industrial waste, the industry must develop technical decomposers by using the processes identified by the circular economy. Thus, the maintenance, reuse, remanufacturing and recycling processes are technical decomposers required by technosphere. The maintenance process has also a role of support by providing the information on the nutrient to regeneration management process, which selects the best decomposer for each nutrient. After the nutrient regeneration, it returns in a firm by being based on flows study of industrial ecology. Nevertheless, to reintegrate a cycle, the nutrient must satisfy many properties tied to the sphere. The nutrient properties (technical and ecological) must be measurable to estimate the impact of nutrient on a sphere. In this way, those properties allow us to characterize the nutrient according to its impact on a sphere and, to determine which decomposers (natural or technical) used.

In this paper, we have identified some directions of research to answer the requirements fixed by the sustainable development and the circular economy. In fact, we have defined the notion of nutrient and determined, among others, that the maintenance process plays an important role in the regeneration paradigm. However, making regeneration in technosphere is not so easy. We must consider the two spheres as a whole. This consideration affects all levels of product lifecycle, from design to retirement. For example, an object will be designed to satisfy of nutrient properties, its regeneration; the control/command will control that the used elements to produce an object will not deteriorate the nutrient properties; the maintenance will keep a nutrient in good condition for the regeneration.

A track for these works is to define the technical properties to determine if a nutrient is regenerable or not and which decomposer is the more adapted to regenerate this nutrient. The manager role must be also defining. In the regeneration context, the nutrient manager must know the value of nutrient properties in real time. With this knowledge, it can identify the nutrient category and determine if the nutrient is regenerable by a technical or a natural decomposer. The manager must monitor the nutrient state of health. The alteration of nutrient properties must be anticipated to avoid changing the type of nutrient. For example, a regenerable nutrient must be regenerated before to become non-regenerable. The new maintenance approach from community Prognostic and Health Management (PHM), which predicts the future state of a system, can be used to determine the time when the nutrient will change state. The regeneration manager must also allow selecting a decomposer depending on the nutrient state, enterprise strategy, costs cause by the different regeneration processes, etc. while avoiding the storage of nutrients. Therefore, by taking inspiration from PHM, the maintenance will be able to supply the needed information of RPM (Regeneration Potential Management).

References

1. European Commission.: Europe 2020 : A Strategy for Smart, Sustainable and Inclusive Growth. Publications Office, Brussels (2010)
2. World Commission on Environment and Development.: Our Common Future (1987)
3. Foundation Ellen MacArthur.: Report Version 1, 2 & 3 : Towards the Circular Economy. http://www.ellenmacarthurfoundation.org/
4. Benyus, J.M.: Biomimicry. HarperCollins, New York (2009)
5. Ayres, R., Ayres, L.W.: A Handbook of Industrial Ecology. Edward Elgar, Cheltenham (2002)
6. Nielsen, S.N.: What has modern ecosystem theory to offer to cleaner production, industrial ecology and society? The views of an ecologist. J. Clean. Prod. **15**, 1639–1653 (2007)
7. Ehrenfeld, J., Gertler, N.: Industrial ecology in practice: the evolution of interdependence at Kalundborg. J. Ind. Ecol. **1**, 67–79 (1997)
8. Du Plessis, C.: Towards a regenerative paradigm for the built environment. Build. Res. Inf. **40**, 7–22 (2012)
9. Cole, R.J.: Transitioning from green to regenerative design. Build. Res. Inf. **40**, 39–53 (2012)
10. McDonough, W., Braungart, M.: Cradle to Cradle: Remaking the Way We Make Things. Farrar, Straus and Giroux, New York (2010)
11. Kormondy, E.J.: Concepts of Ecology. Prentice-Hall, Upper Saddle River (1969)
12. Geng, Y., Côté, R.P.: Scavengers and decomposers in an eco-industrial park. Int. J. Sustain. Dev. World Ecol. **9**, 333–340 (2002)
13. Kumar, S., Putnam, V.: Cradle to cradle: reverse logistics strategies and opportunities across three industry sectors. Int. J. Prod. Econ. **115**, 305–315 (2008)
14. Le Moigne, J.-L.: La théorie du système général: théorie de la modélisation. Presses Universitaires de France (1994) (in French)
15. Mayer, F.: Contribution au génie productique: Application à l'ingénierie pédagogique en atelier inter-établissements de productique lorrain. Phd Henri Poincaré University, Nancy (1995). (in French)
16. ISO 13306.: Maintenance Terminology (2010)
17. Halpin, T.: Object-role modeling (ORM/NIAM). In Handbook on Architectures of Information Systems (Chap. 4) (1998)
18. McDonough Braungart Design Chemistry (MBDC): C2C Certified : Material Health Methodology. http://www.c2ccertified.org/ (2013)

A Case Study of Applying Complexity Leadership Theory in Thales UK

Dawn Gilbert, Laura Shrieves and Mike Yearworth

Abstract Organisations with core capabilities in systems engineering solution development often fail to meet delivery expectations in terms of cost and time-frame. This outcome is viewed as an emergent property of the development organisation, which can be considered a Complex Adaptive System (CAS). The context needed to support complex technical innovation within the organisational CAS appears to be in conflict with a hierarchical bureaucracy in development organisations, whose methods and approaches are best suited to simple and complicated contexts. The paper identifies Complexity Leadership Theory (CLT) as a framework that may offer a way forward in this space. The paper describes two industry-based case studies that sought to practically apply CLT, and provides insights that may be useful to other industrialists interested in applying CLT within their contexts.

1 Introduction

The frequent and high profile failing of systems engineering solutions to be delivered to meet cost and schedule expectations has motivated and shaped a systems research program in Thales UK [1]. Thales UK, like many other businesses

D. Gilbert (✉)
Industrial Doctorate Centre in Systems, University of Bristol,
Merchant Venturers Building, Woodland Road, Bristol BS8 1UB, UK
e-mail: Dawn.Gilbert@bristol.ac.uk

L. Shrieves
Thales UK, Manor Royal, Crawley, West Sussex RH10 9HA, UK
e-mail: Laura.Shrieves@uk.thalesgroup.com

M. Yearworth
Faculty of Engineering, University of Bristol, Queens Building,
University Walk, Bristol BS8 1TR, UK
e-mail: Mike.Yearworth@bristol.ac.uk

G. Auvray et al. (eds.), *Complex Systems Design & Management*,
DOI 10.1007/978-3-319-26109-6_15

with a core practice in development of complex technical systems, has sometimes struggled to meet these delivery expectations. If the organisation is viewed as a CAS [2], with solution delivery as an emergent property, then interventions that attempt to narrow the gap between expectations and delivery must align with a paradigm that appreciates complexity. Interventions in this type of system with this goal in mind can be viewed as an engagement with a wicked problem [3]. Engaging with a wicked problem implies the problems are essentially unique, intervention is a one-shot operation since it is impossible to conduct controlled experiments, solutions are viewed on a spectrum from good to bad rather than purely in terms of success or failure, there is no immediate and ultimate test of a solution, and there is no stopping rule [3]. A case study of an intervention in a wicked and messy problem, therefore, does not produce outcomes that could be thought of as generalisable, since the context and problem are unique. However, the methodology used to intervene in the problem situation can be viewed as generalisable [4]. Therefore, this case study presents research that explores whether something that is possible in theory—the intentional application of Complexity Leadership Theory (interpreted here as methodology)—is also possible in practice.

The overarching Systems Action Research Program within Thales UK engages with this wicked problem, aiming to enhance the ability of its systems engineering function to support solution delivery that meets or exceeds customer expectations. The case study described in this paper was carried out within the Thales UK Systems Engineering Function. Projects within Thales UK are delivered through the transverse and dominant project-led organisational structure [5]. The form of function and project reporting lines are that of a hierarchical bureaucracy. The project-line dominates, and is reductionist in its approaches and methods which suit simple and complicated contexts only [6].

Systems research within Thales UK has previously identified misalignment of methods and problem context as potential contributing factor to poor delivery performance [1]. This issue is not unique to the systems engineering activities within Thales UK. Research by Cowper et al. [7] was based on data gathered in survey responses from 85 systems engineering project professionals. That data set showed that approaches used do not suit the context of the problem, behavior is driven by a narrow view of what an organisation believes is right rather than the broader range of practices they allow, and that project professionals adopt preferred approaches which they which they apply across more than one different type of problem and context (ibid).

1.1 The Problem Situation

The problem situation addressed by this research lies in situations where systems engineering organisations, through mandate or through culture, support the use of approaches (simple or complicated) that do not match the context needed to support productive and innovative systems engineering work (complex), and that systems engineers are not able to specifically identify this misalignment within their practice as part of their decision-making process.

This problem situation resides within the CAS that is the development organisation and an emergent property arising from this problem situation is the gap between what is expected and what is realized in terms of delivery cost and timeframe.

A Possible Way Forward If the organisation is viewed as a CAS and the agents operating within it are currently limited in their ability to identify the nature of their working context, its dynamics as well as which methods and approaches support productive progress, then perhaps an intervention that improves these capabilities may support a transition towards more desirable emergent properties of the CAS itself. Complexity Leadership Theory (CLT) [8] was identified as a conceptual framework that may offer a way forward within this problem space, and may be of interest to similar systems engineering organisations.

2 Complexity Leadership Theory

The CLT is a framework with dynamics [8]. "At its most basic level, Complexity Leadership Theory (CLT) is about leadership *in* and *of* complex adaptive systems" [9, p. 631]. "This framework describes how to enable the learning, creative, and adaptive capacity of complex adaptive systems (CAS)" [8, p. 300]. "This conceptual Framework includes three entangled relationship roles (i.e. adaptive leadership, administrative leadership, and enabling leadership) that reflect a dynamic relationship between the bureaucratic, administrative functions of the organization and the emergent, informal dynamics of complex adaptive systems (CAS)" [8. p. 298]

"Complexity Leadership Theory seeks to foster CAS dynamics while at the same time enabling control structures for coordinating formal organizations and producing outcomes appropriate to the vision and mission of the organization" [8, p. 300]. In 2009, the theory was considered in the context of bureaucratic forms of organising to generate emergence and change in organisations [9].

"The unit of analysis for Complexity Leadership Theory is the CAS. The Boundaries of CAS are variously defined depending on the intent of the researcher, but however identified, they are, without exception, open systems" [8, p. 302].

2.1 Applying Complexity Leadership Research

Brown identifies two strands of complexity leadership research: "There appear to be two general types of research on the behaviors required to engage in complexity leadership. In the first case, some researchers…have identified the principles of complexity sciences and then extrapolated leadership behaviors from them. The second variation consists of researchers…who have longitudinally studied (sometimes retroactively) organisational and inter-organisational emergence phenomenon, using the lens of complexity leadership theory and begun to validate the behaviors predicted by complexity leadership theory. There has been no longitudinal research done to date that I am aware of in which leaders intentionally applied complexity leadership theory to their organisation and overall organisational performance was monitored." [10, pp. 8–9].

The literature search performed in support of this work identified recent examples of empirical research that further developed CLT [11], however no literature describing the intentional application of CLT was found. This case study attempts to intentionally apply CLT.

Considering the intentional application of CLT, Brown identifies a potential limitation regarding the degree of meaning-making maturity that may be required to effectively engage with it. "Experts tend to be immersed in the logic of their own craft and regard it as the only valid way of thinking" [10, p. 18], "the training of it should probably be reserved for leaders who have demonstrated advanced (i.e. post-conventional) meaning-making capacity. It does not seem realistic to expect leaders with a conventional action-logic to learn and sustainably engage with it over an extended duration" [10, p. 18].

3 Research Design

It is appreciated that the case study organisation is dynamic and there are often great demands on the time of systems engineers, which limits the amount of time and attention they have to apply to discretionary activities. Individual systems engineers, however, often have an appetite for accessible and novel concepts that may give them a clearer or deeper understanding of their context, the problem at hand, or methods to help progress work. Appreciating the balance of these pressures, opportunities were sought to introduce CLT concepts to stakeholders within the organisation and to offer resources to support those who wished to learn more or put the theory into practice. Positive uptake is seen as an indication that the theory is initially viewed as promising, however, a lack of uptake within this context is not indicative of the theory being considered impenetrable, irrelevant, unusable or lacking in value.

A case study approach was taken by introducing CLT as expressed by Uhl- Bien and Marion [9] in successive levels of detail to stakeholders within the Thales UK Systems Engineering function. Where initial interest was expressed, further information and research support was provided. This is consistent with an action research based approach of engagement with a wicked problem. This approach was selected with an awareness of the points made by Brown [10]; the meaning-making maturity of the Thales UK stakeholders was unknown, as was the level of meaning-making maturity that would be needed.

3.1 Case Study 1

Initiation The concepts of CLT were initially introduced via an email conversation to the Key Stakeholder in Case Study 1 as a theory considered relevant to an industrial problem that had been explored in an unstructured discussion the week before. This industrial problem related to the apparently conflicting needs of a large organisation which develops large complex technical solutions to control its business through extensive application of reductionist and prescriptive processes, yet also provide the intellectual latitude and freedom needed to develop technical innovations. The email contextualized research relating to complexity in systems engineering development lifecycles, new product development, leadership, business, and management by Akgun et al. [12], Braha and Bar-Yam [13], Hazy and Uhl-Bien [14], Houglum [15], IBM [16], Lichtenstein and Plowman [17], Nugent and Collar [18], Shreiber and Carley [19], Uhl-Bien et al. [8], Uhl-Bien and Marion [9], and Van Oorschot et al. [20] to the problem as it was understood at the time. The Case Study 1 Key Stakeholder was the Thales UK Head of Systems Engineering. In CLT parlance, a role that traditionally was expected to lead by carrying out and overseeing administrative leadership tasks, while also is responsible for successful development and delivery of technical innovations across the organisation. The concepts of CLT described in the initial email underpinned a further one-to-one discussion, which elaborated on the concepts within CLT and how it related to the organisational context at the time.

Method Consideration of CLT within the problem context led to the Key Stakeholder taking an 'extraordinary' step of purposefully enabling a group of 20 Systems Engineering architects, from across a diverse range of Thales UK Domains and Business Lines, to gather together for a week-long workshop to explore and possibly develop a common core architecture for use across all Thales UK business lines. This step was 'extraordinary' within the organisational context at the time, its uniqueness illustrates that the approach was a purposeful application of a novel theory, and not a continuation of business as usual. The Key Stakeholder was able to apply influence to enable presence and participation from a group of Systems Engineers who would normally be under immense pressure to stay 'on project'.

The workshop was held in a design center that supports but doesn't prescribe the use of design-thinking concepts. The normal prescriptive and detailed processes that the architects would generally work within were "banned from the room" (Key Stakeholder, workshop day 1). On Monday morning, at the start of the workshop, the Key Stakeholder provided a brief introduction to participants that described the broad remit and aims for the week, and allowed, in fact encouraged the participants to self-organise. "run it fairly loose, control and process is minimal…we've got tools here, uncontrolled space…use your imagination…have fun, enjoy…you are all intelligent people…its self-organising in the extreme" (Key Stakeholder, workshop day 1).

The Key Stakeholder and workshop participants were aware that the work they were doing was part of ongoing Thales UK systems research. The workshop was video recorded, portions were audio-recorded, several photographs were taken, and field notes were taken by the action-research participant/observer. Participants were invited to submit free-form email feedback on how they had found the experience.

Data The data collected as part of this case study is comprised of: meeting notes; emails; notes taken during phone conversations with the Key Stakeholder in advance of the workshop; video recordings; photographs; audio recordings; field notes gathered during the workshop; and, email-based feedback provided by participants after the workshop.

Selected comments from email feedback include:

- "The approach to the workshop removed normal project/organisational constraints thus enabling the team to realize their potential"
- "In just four days of Design Centre enabled, Cross Domain, Cross Discipline Co-Architecting activities we have achieved what had previously taken (in my experience at least three times now) at least a year if not more. Co-engineering activities have been delivering some successes across the UK now for a couple of years but this activity has in my "humble" opinion pinnacled them all."
- "We all have our own mental models of [the core architecture] and much of that is shaped by past experience but despite that we were all able to think outside the box and that is fundamental to the success of the event….we made more progress over 4 days than we had in the previous two years with the one day workshops we ran"
- "When I first entered the Design Centre, I was taken back by its informal nature —but having experienced it I must say it works….Where we got in the four days (and a half) was quite an achievement…overall and excellent experience."
- "I have not seen any initiative in Thales that has been as dynamic, constructive and productive or achieve the level of cooperation and cohesion within a team that covered many disciplines and business lines"
- "Outstanding opportunity taken to get the right people in the right place for long enough to make real forward progress on a critical transverse topic that can enable business effectiveness in the long term…there is a key action to determine and sell the value proposition for not only the [core architecture] approach, but also the process of collaborative exploration that we have followed this week".

3.2 *Case Study 2*

Initiation The concepts of CLT were initially introduced to one of the case study 2 stakeholders by copying them on an email to a different audience, that focused on a different topic, but which referenced and included Uhl-Bien et al. [8] as one of many attachments. This introduction to CLT suggested, to the stakeholder, that CLT might hold some promise towards addressing an issue they had recently discussed with another case study 2 stakeholder. Further discussion between these two Key Stakeholders led to a request for broader and deeper engagement by the research team.

Method A group of systems engineers based on the same site as the two Key Stakeholders were invited to a 90 min briefing session on CLT. Those who couldn't make the originally scheduled session were invited to participate in a second session, which was held around 10 days later.

The term 'systems engineer' in Thales UK covers a broad spectrum of role types, as may be expected from an organisation spanning diverse operating domains and solution types. Similarly, depending on the programme, systems engineers may varying levels of involvement with customers, project managers, systems engineering peers, engineers within other specialisms, subcontractors, and colleagues specializing in areas such as quality or purchasing.

As participants entered the CLT briefing session they were advised that the session was being used for research purposes and that, as such, the discussion was being audio recorded. They were asked to complete a single-sided A4 hard copy 'before' survey which was gathered back in before the briefing commenced. This survey was designed to ask non-leading questions to establish the potential relevance of the CLT concepts for the role each participant performs as well as to gather initial views on their appreciation of and perceptions relating to socio-technical complexity. A briefing was then given which briefly described the Cynefin Framework (Snowden and Boone [6], then built on that description to introduce CLT. The Cynefin Framework was introduced initially as a basis to provide a tangible definition of complexity, and to introduce how different contexts suit different approaches to progress. It has been described to other audiences within Thales UK's Systems Engineering Function before, and found to be a description of complexity that can be appreciated quickly. The briefing also mentioned that the UK Head of Systems Engineering had put CLT into practice successfully. After the briefing, a second single-side A4 hard copy survey was handed out which asked participants to rate on likert scales the prevalence of contexts (as described in Snowden & Boone, [6] in their work environment, and their use of CLT behaviors (as described in [9].

The participants were then invited as a group to discuss their initial views of the frameworks and to discuss whether they thought these related to their own work. Before departing participants were invited to note on the back of their 'after' surveys 2 or 3 opportunities that would occur within their normal work in the upcoming month where they could consider the frameworks in advance of, during and after the work. Participants were advised that after the opportunities occurred they would be briefly interviewed by the research team (in person or by phone, depending on what was most convenient) to gather their views on how relevant the frameworks were 'in real time' and to see whether reflecting on their initial introduction to these frameworks was able to support their everyday decision-making. Participants were advised that this would likely take 10–15 min.

The annotated 'after' surveys (which could be linked to an individuals 'before' survey) were collected, converted to an anonymized soft copy, and emailed back to each participant, thanking them for participating, giving them initial feedback on how their surveys responses may be interpreted, and suggesting a time and method for gathering reflections on the attempted application of theory.

In all, 16 participants took part in the two briefing sessions and completed the surveys (see Table 1 for a summary), 14 participants agreed to consider the frameworks in their work. Three participants couldn't be reached for feedback on their practical application; however feedback was gathered from the remaining 11 participants by a one-to-one in-person or telephone-based semi-structured interview. The semi-structured interviews used a tone and vernacular that mirrored the participant. Questions explored how participants understood the frameworks and how they related them to their environment, and their own behavior and the behavior of their colleagues. The author carried out all the interviews over the course of 2.5 weeks to ensure consistency. One of the interview summaries was provided back to the interviewee for comments to confirm whether this note-taking approach was able to accurately capture content and intent of the discussions. The interviewee stated no editing was required.

Data The dataset for this case study, therefore, includes the email exchanges and notes from phone calls which led to the briefings being held, the before and after surveys, the presentation materials and audio recordings of the briefings and follow-up free form discussion, one-to-one email exchanges regarding interpretation of the surveys and opportunities to apply the frameworks, and the interview

Table 1 Summary of selected 'After' survey responses

How often do you observe these contexts in your work?					
Context	Never	Rarely	Sometimes	Often	Always
Simple	0	1	4	6	5
Complicated	0	0	5	11	0
Complex	0	1	8	7	0
Chaotic	0	7	7	2	0
How often do you engage in these leadership behaviors?					
Behavior	Never	Rarely	Sometimes	Often	Always
Administrative	0	3	1	12	0
Enabling	1	1	10	4	0
Adaptive	2	2	8	4	0

notes which were promptly written up electronically based on hand-written notes made during the interviews.

Selected quotes from the follow-up interviews include:

- Participant 1: "I do a different job, front-end, dealing with sales and marketing… from and engineering perspective….it's a mature, repetitive process"
- Participant 2: "I think in my case you have your own style which is the way you manage projects. If you start thinking of contexts, then you can select methods that work and if you think about it and select the right methods, that becomes your new style…Another thing that might be interesting to explore - this is targeting engineering, but things like QA [Quality Assurance] and purchasing perhaps could do with more CLT work with these functions.We need more experience in admin, enabling, adaptive. We need a bit more time to recognize and learn how to act"
- Participant 3: "It was definitely in chaos.it was a rollercoaster…trying to follow the ideas you presented was very difficult…there were occasions where in particular I could see contexts…but you have limited influence, not none, you can always do something…list the assumptions you used…Others were making decisions…That said, I've never seen anything quite this bad…In response to the question you asked which was did it have any use, I'd have to say 'limited'…I would recognize areas of the business where the bureaucracy is more restrictive than it needs to be to provide support for engineers or anyone to use their initiative or take responsibility to come up with their own ideas, I can see that, but it's not black and white….if its right for the present, is it right for the future? That becomes a difficult way of looking at it, whoever looks at it has their bias"
- Participant 4: "What you were saying…I can relate to it, being adaptive…it was really good to listen to you, it makes perfect sense, it was good to be on a project that is practicing the approach…In a different team there are different ways… about 2 years ago, I wouldn't necessarily be able to say that it would have been relevant, but for me at least the timing of your lecture was perfect"
- Participant 5: "I can clearly see it [CLT] applied to the business we are in…I certainly had no problem understanding how it related…the problem I've got is in doing something…I'd say that's the enabling part - money and support. People and commitment are lacking, they say 'that's a good idea get on with it', but then you can't"
- Participant 8: "We've just been too pushed so far…I haven't looked at the frameworks"
- Participant 10: "I want a single page that tells you what to do to do your job"
- Participant 12: "Throughout the [first] meeting I was aware of the different sorts of leadership behaviors and I could adjust. I was more aware of the styles, but I'm not sure whether it changed what I did. I probably would have done the same thing, but maybe the clarity helped me to do it earlier……The [second] meeting itself wasn't productive, but afterwards S and I spent about 2 h on chairs in the open area discussing how to get value out of the group, and it was

amazing…It was a weird one, in the meeting I'd wanted to do enabling, but they don't understand the problem enough, so I had to do administrative behaviors. With S we were adaptive in how we came up with ideas about what to do, it was great….If S and I hadn't got so disenchanted with how the meeting went I'm not sure the later one [in the open area] would have happened."

- Participant 14: I find myself flitting about the three behaviors, administrative, adaptive, enabling all the time…the framework keeps you sane, you need something to help you navigate when you have to flit around…I don't feel constrained by the SEM [Systems Engineering Manager] role, I look at the processes as providing a good guideline of what is needed, what needs to be established for quality etc., but it doesn't tell you how to work, that's up to you. I mean your frameworks…I recognized it all…in summary I think the frameworks don't tell you how to do things, you can say here's things I recognize, here are some pointers on what to do"
- Participant 15: "I definitely recognize them [the three CLT behaviors], I don't get to do anything with them…though, we are so busy….There is a preference to serve issues that are short-term…we plan so many programs as if they are simple, and on almost every program there's something that gives us a problem, that's not deterministic…If there's a big problem of course the first question asked is how long will it take to fix and how much will it cost, but the people solving it don't know, then it becomes a bit chaotic, you have to do the work to find out, it takes as long as it takes, people are always pushing specific plans and timeframes…its more the PM [Project Management] world, and I can understand where they are coming from, they don't like unknowns…For PMs the more we can make them understand the problem the better, it doesn't feel like they do. If you get them in a room and they understand it, they'd struggle to recognize it when they faced it in their work the next day".

4 Analysis

The data collected during the course of the two interventions was reviewed with the aim of answering the compounding questions of whether CLT is understandable to practicing systems engineers? Are the components and dynamics described by CLT recognizable in the environments those systems engineers work in? Is it possible to apply CLT within the Thales UK systems engineering context? Can applying CLT within the Thales UK systems engineering context be valuable?

The evidence from these case studies supports the view that CLT can be understood by practicing systems engineers. The Key Stakeholder in Case Study 1 was able to take specific actions on the basis of appreciating the theory and how it relates to the problem context. The survey responses in Case Study 2, presented in Table 1 suggest that respondents could both understand and relate CLT behaviors (as well as the Cynefin Framework) to their work environment and actions. The ability of the Case Study 1 Key Stakeholder to deliver the core architecture workshop demonstrates that CLT can be applied within the Thales UK systems engineering context, as do the comments provided by Case Study 2 Participant 12. The feedback comments from Case Study 1 workshop attendees demonstrates that the application of CLT which led to the workshop being held, and guided how it was run supported achievements made during the workshop that are considered valuable within Thales, and unachievable using 'normal' approaches. Case Study 2 Participant 12 wasn't sure whether the application of CLT led to better decisions and actions, although they note that they may have made their decision more quickly, which supports the notion that the further application of CLT within Thales UK could provide value. Case Study 2, Participant 14, reported that they use each of the CLT behaviors and whilst knowledge of the theory itself may not change what they do, familiarity with the framework provides a valuable structure from which to navigate from.

5 Discussion

The case studies performed within Thales UK demonstrate that CLT can be intentionally applied, as demonstrated in this systems engineering context. It's application can support the achievement of desirable outcomes more quickly and with greater confidence. The data collected also suggests that outcomes that could not be achieved within 'normal' operating conditions may be possible via the intentional application of CLT.

Brown [10] proposes that leaders with more mature meaning-making systems may be more capable of engaging with practices of complexity leadership. Conversely those with conventional meaning-making systems may not be able to fully adapt to the fundamental changes in leadership perspectives called for by complexity leadership. The analysis of the two case studies reported here show there was variety in the need and inclination of participants to recognize the elements and dynamics of CLT within their own context. Participants 1 and 10 of Case Study 2 gave the three responses stating that they never used enabling or adaptive leadership behaviors. Given the description participant 1 gave of their role, there is no indication that this contributes to undesirable emergent properties of the CAS, however viewing feedback from participant 10 in the round, they appear to have struggled to appreciate the relevance of these concepts to their role and environment. It could be the case that, as a population, systems engineers are predisposed to better recognize how CLT relates to their work, since complexity is a core feature of much of their work. Participants 2 and 15 noted that other Thales UK functions could benefit from understanding CLT, though Brown's view is supported by Participant 15, in claiming that the Project Management community would struggle to apply the theory in practice.

A variety of comments related to time. Participants 8 and 15 indicated they were too busy to consider or use the frameworks. Under the pressure of limited time, participants 3 and 15 identified the apparent tension between interests in pursuing short-term and longer-term value. Participant 4 had, within the last year, transitioned into a new business line after more than 20 years in a different Thales UK business line. They clearly identified CLT in their actions transitioning into and leading the systems engineering in a new technical area and found the concepts valuable, although they noted that prior to the move into a new work environment the theory may not have seemed relevant. Brown [10] identifies that managers may not be able to sustainably engage in complexity leadership without regular support, however, participant 2 states that in their case, improved abilities to understand context and act appropriately becomes embedded in normal practice.

6 Conclusion

This research has provided an example of how CLT can be introduced to agents within a CAS as a means to intervene in a wicked problem. A variety of short-run responses were observed and are included in this paper, which range from immediate comprehension leading to an application which realized high- value outcomes in the short-term (case study 1), to an apparent inability to see how CLT applied to the working environment (case study 2, Participant 10). These results demonstrate that it is possible to intentionally apply CLT in practice. Longer-term impacts from these interventions continue to be felt within Thales UK, as would be expected from an intervention in a wicked problem.

Acknowledgments Dawn Gilbert is registered on the Engineering Doctorate (EngD) Program in Systems at the University of Bristol. This work is supported by the University of Bristol, UK Systems Centre, the EPSRC funded Industrial Doctorate Centre in Systems (Grant EP/G037353/1) and Thales UK.

References

1. Gilbert, D., Yearworth, M., Oliver, L.: A Systems approach to the development and application of technical metrics to systems engineering projects across an enterprise. In: 12th Annual Conference on Systems Engineering Research (2014)
2. Cilliers, P.: Complexity and Postmodernism Understanding Complex Systems. Routledge, London (1998)
3. Rittel, H., Webber, M.: Dilemmas in a general theory of planning. Policy Sci. **4**, 155–169 (1973)
4. Yearworth, M., White, L.: The non-codified use of problem structuring methods and the need for a generic constitutive definition. Eur. J. Oper. Res. **237**, 932–945 (2014)
5. Hobday, M.: The Project-based organisation: an ideal form for managing complex products and systems? Res. Policy **29**, 871–893 (2000)
6. Snowden, D., Boone, M.: A leaders framework for decision-making. Harvard Bus. Rev. **85**, 68–76 (2007)

7. Cowper, D., Elphick, J., Kemp, D., Evans, R.: To V of not to V—That MUST be the question knowing when to apply the right approach. In: Proceedings of INCOSE International Symposium, vol. 24(1), pp. 793–814 (2014)
8. Uhl-Bien, M., Marion, R., McKelvey, B.: Complexity leadership theory: shifting leadership from the industrial age to the knowledge era. Leadersh. Quart. **18**(4), 298–318 (2007)
9. Uhl-Bien, M., Marion, R.: Complexity leadership in bureaucratic forms of organizing: a meso model. Leadersh. Quart. **20**, 631–650 (2009)
10. Brown, B.: Complexity Leadership: An Overview and Key Limitations (2011). http://integralleadershipreview.com/3962-learner-paper-complexity-leadership/ (14 April 2015)
11. Havermans, L.: Leadership in project-based organisations: dealing with complex and paradoxical demands. PhD dissertation, VU University, Amsterdam (2014)
12. Akgun, A., Keskin, H., Byrne, H., Ilhan, O.: Complex adaptive system mechanisms, adaptive management practices, and firm product innovativeness. R&D Manage. **44**, 18–41 (2013)
13. Braha, D., Bar-Yam, Y.: The statistical Mechanics of complex product development: empirical and analytical results. Manage. Sci. **53**(7), 1127–1145 (2007)
14. Hazy, J., Uhl-Bien, M.: Towards operationalizing complexity leadership: how generative, administrative, and community-building leadership practices enact organizational outcomes. Leadership **11**(1), 79–104 (2013)
15. Houglum, D.: Myth-busters: traditional and emergent leadership. Emergence: Complexity and Organisation, VOL. 14(2), pp. 25–39 (2012)
16. IBM: Capitalizing Complexity, Insights from the Global Chief Executive Officer Study, (2010). http://www-935.ibm.com/services/us/ceo/ceostudy2010/ (15 April 2015)
17. Lichtenstein, B., Plowman, D.: The leadership of emergence: a complex systems leadership theory of emergence at successive organizational levels. Leadersh. Quart. **20**, 617–630 (2009)
18. Nugent, P., Collar Jr., E.: The hidden perils of addressing complexity with formal process—a philosophical and empirical analysis. In: Proceedings of the Fifth International Conference on Complex Systems Design and Management, pp. 119–131. Springer, Berlin (2014)
19. Schreiber, C., Carley, K.: Leadership style and an enabler of organizational complex functioning. Emergence: Complexity and Organization, vol. 8(4), pp. 61–76 (2006)
20. Van Oorschot, K., Sengupta, K., Akkermans, H., Van Wassenhove, L.: Get fat fast: surviving Stage-Gate® in NPD. J. Prod. Innov. Manage. **27**(6), 828–839 (2010)

A NAF-Based Proposition to Leverage System Engineering Change Management in Systems-of-Systems Acquisition Project Teams

Thomas Rigaut

Abstract A key issue in systems-of-systems acquisition agencies is enforcing effective use of System Engineering at the scale of project teams. Usual change management strategies would not bypass the high cost of architecture modelling at the scale of a system. This article proposes a pragmatic approach to minimize this investment by gradually incorporating engineering works from specific domains, in order to constitute a NAF-based technical-oriented referential. Use cases and a case study highlight how such a referential can be exploited to fuel technical analyses, then the decision-making process of the acquisition project, and thus providing incentive to the team to get leverage for System Engineering change management.

1 Introduction

Most industrial organizations aim to bolster use of Systems Engineering by their project teams, because of the evidence of concrete gains obtained by System Engineering application. Moreover, conducting change management operations inside large organizations is increasingly common in the context of a worldwide competitive economy.

However, this article states that managing change of engineering practices into enforcing effective use of System Engineering raises specific issues at the scale of project teams. This article focuses on project teams involved in complex defense systems or defense systems-of-systems acquisition projects, constituted under matrix organization: the decisional part of team consists in people coping with Project Management issues and decision-making; the technical part of the team copes with specific Engineering Issues (domain specialists), and pilots System Engineering processes encompassing the whole technical activities.

T. Rigaut (✉)
Direction Générale de l'Armement, 7 Rue des Mathurins, 92221 Bagneux Cedex, France
e-mail: thomas.rigaut@intradef.gouv.fr

G. Auvray et al. (eds.), *Complex Systems Design & Management*,
DOI 10.1007/978-3-319-26109-6_16

Every approach for bolstering use of System Engineering in Project teams shall address a redefinition of the technical processes of the organization based on a State-of-the-art standard, such as ISO 15288 or IEEE 1220-2005; then enforcing use of adapted Requirements Engineering and Architecture Modelling methods, supported by efficient tools.

Personal experience of the author suggests that redefining technical processes proves difficult until feedback on System Engineering application has been collected among some pioneer teams of the organization. Establishing Requirements Engineering practices quickly gives added value with low investment: concepts are easy to apprehend such as traceability, or engineering documents data models, and mature engineering tools are available on the market.

Whereas Architecture Modelling is founded on a largest variety of concepts, moreover depending on the choice of the Architecting Framework, and supported by uneven architecting tools: therefore the investment needed to be able to gather the key engineering data in a ready-to-use consistent architecting model would seem too high for engineers who perform domain specific analyses such as "Analyzing the dysfunctional modes of a subsystem", "Analyzing the performance of a functional chain", "Optimizing an hybrid Hardware/Software System", "Trading-off between quality requirements and performances".

Therefore a key to manage System Engineering changes is providing incentives to the technical team members into delaying their analyses under work for contributing to the construction of a mutual architecting model, able to provide consistent and validated data to the whole team, therefore providing data for system-level or system-of-systems-level analyses.

The author experimented three approaches and experienced their limits. First, asking domain specialists to delay their activities in order to invest time into building a common engineering referential: it only works at the very beginning of projects, where capitalizing existing data is costless. Second, assigning one architecture-modelling skilled member of the technical team the task to reassemble the disseminated data into one consistent model: the model is little exploited because it is not updated frequently enough nor directly responding to day-to-day preoccupations of the technical team. Third, defining modelling objectives touching subset of the technical teams, then model architectures toward these objectives. A successful method can be found in [1], but it does not converge towards a complete repository of key engineering data at the scale of the project.

This article develops an approach to minimize investment of technical team members for creating a NAF-based technical-oriented referential at the scale of the project, able to provide added value to domain specialists, and addressing specific systems-of-systems issues: functional chains analysis, network definition, multi-objective optimization [2].

To show this alternative viable, this article first presents the earned value expected from using NATO Architecture Framework into our approach. Then this article describes the steps of the approach. Some use cases will outline how value is earned through the approach and leverage on change created, while a NAF-based technical referential is constituted. Finally a case study is presented and advocates for delivering a straightforward vision of System Engineering role into a Systems-of-Systems or complex System acquisition project.

2 Earned Value Expected from Using the NATO Architecture Framework (NAF) into Our Approach

NATO Architecture Framework is an architecture framework particularly fitted to describe defense systems-of-systems [3].

NAF 3.1 is based on the NAF Metamodel (NMM) [4]. This Metamodel defines concepts covering much of the issues raised by stakeholders of defense systems-of-systems; so that a NAF-based technical-oriented referential can meet the concepts used by most of the stakeholders of the acquisition project.

NAF provides for complementary viewpoints, so it is easy to give feedback to stakeholders from a NAF-based technical-oriented referential, through relevant NAF Views, each addressing a specific issue.

The NMM provides ways to ensure consistency between these viewpoints, so that the consistency of a NAF-based technical-oriented referential can be analyzed and improved whenever new data is collected.

NAF is implemented by profiles of market tools such as System Architect or MEGA Suite for NAF.

In order to cover our assumptions, let's present the NAF viewpoints and the various issues they address:

- The NCV (NATO capability viewpoint) and NPV (NATO program viewpoint) both allow to describe the high-level needs of an organization, like the set of capabilities a modern army needs, as well as the acquisition strategy adopted by the organization to deploy the successive capabilities increments. Each acquisition project led by the organization refers to a subset of the needed capabilities —often described in a national white paper—and ought to follow the acquisition strategy—as determined by defense planning laws.
- The NOV (NATO operational viewpoint) and the NSOV (NATO service-oriented viewpoint) both allow to describe explicitly how the system would work in all its considered systems-of-systems contexts. These viewpoints may focus on what information is exchanged between systems (it is the NOV way) or what service agreement between the component systems would make the systems-of-systems federation work (it is the NSOV way). By describing the way the system would make a system-of-systems run, an architect would also describe how the system would contribute to the capabilities needed by the acquiring organization.
- The NSV (NATO System Viewpoint) where alternative solutions for implementing the operational need (depicted by NOV and NSOV) are described. It is the place where all domain-specific data is to be filed, whether coming from the

NPV-NCV aggregate	Building a capabilities referential and an acquisition planning.
NOV-NSOV aggregate	Building an operational need referential relevant to various systems-of-systems contexts.
NSV-NTV aggregate	Building a technical referential tracing all contemplated alternative solutions to operational need.
NAV and bridging views	Dynamically ensuring consistency between the three former aggregates, thus progressively comforting one technical-oriented referential built on three pillars. Organizing engineering work around the referential.

Fig. 1 Presentation of NAF aggregates

technical team or industrial stakeholders—current or to-be holder of developments contracts. The NSV has a strong adherence with the NATO Technical Viewpoint which focuses on all technological and normative constraints—and opportunities.

- The NAV (NATO All Viewpoint) where the technical team describes its System Engineering strategy (NAV-1), a shared glossary (NAV-2), a common architecting data model (NAV-3b) and modelling rules to enforce (NAV-3a). To this viewpoint we associate all the bridging views of the NAF, e.g. the views such as NSV-5 which links operational activities from NOV-5 to system functions form NSV-4 (for those unaccustomed to NAF, some bridges remain implicit, for instance NOV-2 Operational Nodes to NSV-1 System blocks bridge.

From this description of four NAF aggregates, we can propose a mapping to four core activities that lead to the building of a consistent NAF-based technical referential for a complex system involved in system-of-systems contexts (Fig. 1).

3 Description of the Approach

Our approach follows four steps: incorporating data from domain specific works while ensuring formal consistency of collected data; performing architectural analyses and providing deliverables to initial contributors through NAF views; building synergies between engineering processes through the NAF-based technical-oriented referential; providing global analyses results to decision-makers.

3.1 Step 1: Incorporating Data from Domain Specific Works

Incorporating domain-specific data to a NAF-based technical-oriented referential requires a preliminary work inside the technical team of the project: consistency

NAF Metamodel concept	Proposed interpretation	Usefulness
Capability measure of effectiveness	High-level functional objective	Objective function for optimization process
	High-level quality objective	Objective function for optimization process
	High-level cost objective	Constraint function for optimization process
Project	Acquisition project	Evaluate impact of unsynchronized acquisition processes of the component systems of the System-of-system
Capability	Capability	Capability an organization is willing to acquire
Capability Increment	Capability Increment	Increment acquired through one acquisition project
Capability Configuration	System deployable configuration	Definition of the system that provides the desired capability, so is aimed for deployment

Fig. 2 Proposed interpretation of NPV-NCV concepts

between NAF concepts and those of specific engineering activities should be evaluated, so that specific concepts would be mapped on NAF concepts, or extensions to the NAF Metamodel (NMM) would be considered.

Concept mapping between specific activities can prove difficult to establish, that's why we have to resort to NAF Metamodel as a pivot between concepts named after their role in specific engineering activities. This article proposes to focus on a simplified vision of the core of NAF Metamodel, completed by a few key concepts from requirements engineering (Figs. 2, 3 and 4).

This NAF Metamodel interpretation comes from field experience of presenting NAF concepts to technical project teams. It intends to rename concepts in order to conceal with more classical Metamodel such as SysML or notation provided by IDEF0 (centered on the concept of function) or BPMN (centered on the construction of chains of functions supported by actors). This interpretation should allow to better introduce the duality between Operational-Service aggregate, where functions and actors express needs, and System-Technical aggregate, where functions and resources express solution alternatives.

The adjunction of attributes expressing performance expected from or reachable by systems allows to consolidate key specification data into the NAF referential, instead of tracing it from heavier, project management-oriented documents such as System Specifications.

If concepts from specific engineering activities cannot be mapped to one of the concepts of the upper tables, one should try to express this concept as an attribute of one of the core NAF concepts, then try to map it to one of the other NAF concepts

NAF Metamodel concept	Proposed interpretation	Usefulness
Operational Scenario measure of effectiveness	Operational-level functional objective	High-level objective function derived to the operational level, useful for optimization process
Operational Scenario	Operational scenario	A situation where a considered system (or SoS) is expected to fulfill an operational effect through collaboration of its system components
Operational Node	Operational actor	Every actor involved in an operational scenario
Operational Activity	Operational function (Expresses an operational need)	Every function an operational actor has to perform in order to collaborate to an operational scenario
Information Exchange Need	Information Exchange Need (Needline in BPMN)	A set of operational information two operational actor have to exchange in order to perform an operational scenario
Service	Service (Expresses an operational need)	Every function which is expected to be provided through a service agreement, in order to support a system capability
Service agreement	Service agreement	Set of conditions under which a service should be provided, and at what level it is to be provided
	Desired performance	Desired performance of an operational function or of a service

Fig. 3 Proposed interpretation of NOV-NSOV concepts

as described in NAF v3.1 Chap. 5, otherwise importing this concept into the NAF-based referential, and mention it in the NAV-3b NAF Metamodel Extension view.

The method described in [3] would help expressing and organizing concepts for a specific engineering activity, then mapping them to NAF.

Before analyzing consistency from the substantial point of view, one shall fuse redundant data, and harmonize formats and graphical representations.

When two subset of the technical team, leading complementary analysis, are about to produce redundant data in parallel technical processes, Systems Engineering management should provide guidelines either for fusing the produced data in the architecture modelling referential, or for making the two sub-teams cooperate into producing common data for their respective needs, then completing it with their specific data.

NAF Metamodel concept	Proposed interpretation	Usefulness
Architectural resource	Alternative solution	A block considered for implementing an operational node, depending on its reachable performance and quality requirements, plus the technical constraints it brings to the design
Technical function measure of performance	Reachable performance	Performance reachable by a technical function
	Technical constraint	Constraint on the design justified by implementation issues
	Reachable Quality requirement	Quality property of a system resource deemed reachable by the system designer
Architectural resource	Technical architecture Block	A block implementing an operational node
Human resource	Human resource	A human resource part of an architectural resource
Material resource	Material component	A material resource part of an architectural resource or another material resource
Software resource	Software component	A software resource part of a material resource of another software resource
Communication channel	Communication channel	A channel carrying information between two system resources
Communication protocol	Communication protocol	A protocol under which information is carried within a communication channel
System function	Technical function	Function carried out by a material or software component, or a technical architecture Block
System function	Crew function	A function carried out by a human resource
Standard	Technical standard	In the context of NSV only technical standards are considered. They bring design constraints

Fig. 4 Proposed interpretation of NSV-NTV concepts

The three pillars of a System Engineering approach: processes, methods, tools, recall us that even when different technical processes are coordinated, methods are using common concepts, a lack of compatible tools may condemn the approach.

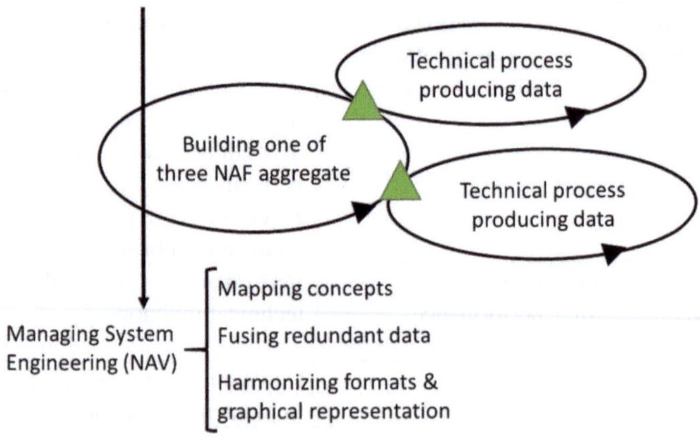

Fig. 5 Ability to capitalize data depends on System Engineering management process

Indeed the NAF technical referential is a database organized through the NAF Metamodel implemented by a computer tool, thus using particular graphical notations. Whether incompatible notations and formats where used, data exchanges would be technically difficult.

To avoid these incompatibilities, the technical team should work out importing formats and scripts able to automatically import new data into the NAF referential, then fusing it to existing data with conflict detection. If the technical team has not such formats and scripts, the NAF referential can be split up between a database carrying lists of concepts and their traceability links, and a database carrying graphical representations.

A second point to consider is enforcing a reduced number of graphical representations types. For instance, scenarios should be described in one executable graphical representation (like Business Process Modelling Notation), should they be simulated afterwards (Fig. 5).

3.2 *Step 2: Building Synergies Between Engineering Processes Through the NAF-Based Technical Oriented Referential*

At a certain point of the project, the technical team manager would decide that enough data has been produced and capitalized into the NAF-based referential, so that the referential would serve as a reference for all the specific technical processes. From that point, all data would be managed in configuration.

This point of the project is also the first occasion to give a feedback to all those who contributed to fill the referential: consistency, completeness of data can be evaluated from an architecture modelling point of view; holes in the architecture are

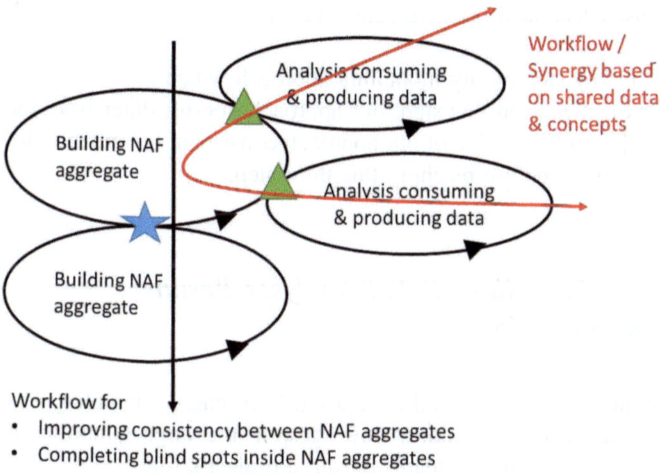

Fig. 6 Corrected engineering workflows, consequence of feedback

identified so that the technical team would focus on the blind spots of the need expression and of the design; data that couldn't be fused indicates sectors of the design that needs to trade between the objectives of the different technical processes using the data off.

After this first feedback, the technical manager can identify synergies to be developed between two or several processes that base their analysis on the same concepts and data, shared through the common referential.

The point of this article, is that an engineering review based on the common referential would easily provide guidelines to develop better, System Engineering-inspired workflows for the technical team: workflows to strengthen the common referential by filling the gaps; workflows to develop new synergies between the different activities led by the technical team—with or without the other stakeholders (Fig. 6).

3.3 *Step 3: Performing Architectural Analyses and Providing Deliverables to Initial Contributors Through NAF Views*

Once enough data is gathered, and synergies developed between specific engineering activities, deliverables shall be defined based on NAF views. These deliverables may address domain specific issues as specific Systems-of-Systems issues. Relevance of these deliverables highly depends on the right use of NAF concepts and their links, and the completeness of the referential. At this step, some specific engineering activities can be redefined by the data they shall produce into the NAF-based technical-oriented referential, and the analysis of the NAF-based deliverables.

This is the key step of any modelling approach, when the contributors get their return on investment. On this step, our approach doesn't differ from the approach described in [3]. Added value of our approach consists in minimizing the entry cost for the contributors before reaching this third step.

3.4 Step 4: Providing Global Analyses Results to Decision-Makers

This article intends to show that the proposed NAF-based referential is well-fitted to harbor a unified decision-making flow, despite the large quantities of analyses results produced by technical teams during a project's life.

The idea is to use the NAF-based technical-oriented referential as a pivot for evaluating the high-level objectives (or objective functions): cost, capability measure of effectiveness, quality requirements.

The technical team would define a performance evaluation model for each NAF aggregate considered, and each objective function; these performance evaluation models are based on architectural properties, and built on a multi-level principle, so that for each aggregate and each objective function, a performance tree can be defined through the different levels architecture.

The nodes of the tree are evaluated through architectural properties—connectivity of components, choice of components, cardinality of components—based on the corresponding level of design and the performances of the lower nodes and leaves of the tree. The leaves of the performance tree cannot be evaluated through architectural properties but are to be evaluated under the rules of specific engineering domains. In other words, this article suggests to build explicit performance models that segregates architecture-based evaluations, and domain-specific-based evaluation.

For one NAF aggregate, once the performance evaluation models are defined for each key objective functions, each alternative design should be evaluated. These alternative designs are the direct result of the data collect from specific engineering activities: these activities focus each on different subset of desired properties, so that alternative designs would naturally emerge in the common referential. It has been mentioned in this article that a fusing process should be discussed within the team: identifying alternatives instead of "mixing" architectures is a natural outcome of such processes.

The result of the evaluation is a table with alternatives as rows, key function objectives as columns, and evaluation results in the cells. These evaluation tables can be produced for every node of the performance tree, not only its root node, so that evaluation can focus on the desired subset of the design. The next question is:

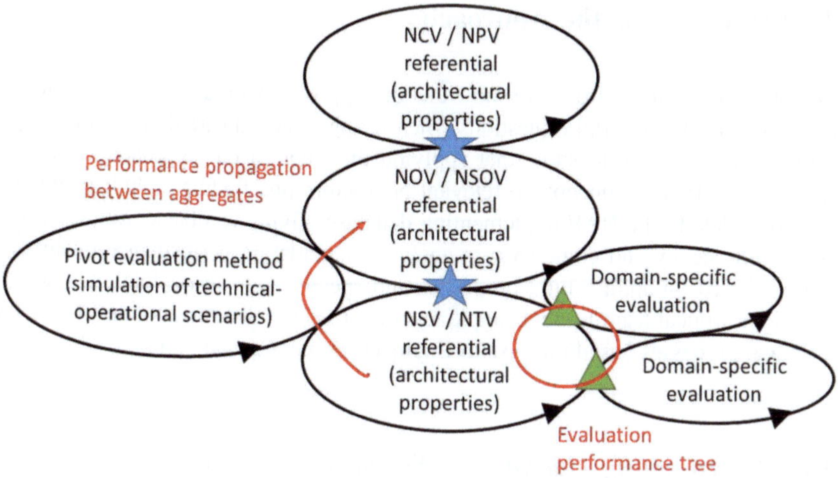

Fig. 7 Evaluation of performances through trees and pivot evaluation methods

what does the team do with this table? It should be considered only as an input to a decision-making process led by project management and client delegates.

In the NSV/NTV case, this process can lead to choose one alternative solution over the others, to identify design guidelines, or to pinpoint performance or quality requirements trade-offs that should reach a better balance.

In the NOV/NSOV case, this process can lead to discard operational concepts, identify operational use guidelines, or to pinpoint to-be optimized trade-offs between different operational scenarios measures of effectiveness, or between the different measures of effectiveness of one operational scenario.

In the NCV/NPV case, this process would lead to investment guidelines: which capabilities shall be prioritized for funding, or for prioritized deployment? On which quality requirement or key performance technological investments should focus?

One key challenge of this approach is the coordination of the evaluations on the different aggregates; it seems rather ambitious to define generic performance evaluation model through multiple aggregates. A pragmatic approach would be to rest on specific engineering analyses using data from two different aggregates: for example, simulating measure of effectiveness of technical-operational scenarios, e.g. scenarios embedding technical data such as sub-system functional performances (from a NAF point of view, an operational scenario supports only expression of operational needs, and therefore operational actors carry desired performance; a technical-operational scenario replaces these operational actor by system resources carrying reachable performances) (Fig. 7).

4 Use Cases for the Approach

The purpose of this section is to show that our approach would lead technical teams to ask themselves the right questions when trying to coordinate their various engineering activity; then to show that applying the steps of our approach—mapping specific-to-activities concepts to relevant NAF concepts, finding technical solution to import data to the NAF implementing database, giving feedback from the referential consistency and completeness analyses, elucidating performance evaluation models based on architectural properties and specific-domain evaluations—would give the technical team leaders simple leverage to get their teams commit to formal technical processes based on System Engineering recommendations.

4.1 Coordinating Operational Scenario Analysis and Functional Analysis

This table shows that not every concepts could be mapped, pointing out that each method can provide specific added value to engineering works (Fig. 8).

Functional analysis allows to easily brainstorm operational functions; these functions are contextualized into operational scenarios, so it becomes far easier to sort these functions.

Performance justification is provided from operational scenario analysis: a performance is justified when it is attested that one operational scenario needs the operational function to reach that performance to attain the objective operational effectiveness.

Using this approach gave the author leverage to lead technical teams from functional analysis methods to modeling simple operational scenarios, thus building a more consistent operational referential than static functional catalogs functional analysis usually provides (Fig. 9).

NAF referential concept	Functional analysis concept	Operational scenario analysis concept
Operational scenario	Life-cycle situation	Operational scenario
Operational function (expressing need)	Function	Operational activity
Operational actor	Interactor	Operational actor
Information exchange need		Needline
Desired performance	Performance	Performance
Operational measure of effectiveness	Assessment criterion	Operational measure of effectiveness

Fig. 8 Concept mapping for functional analysis

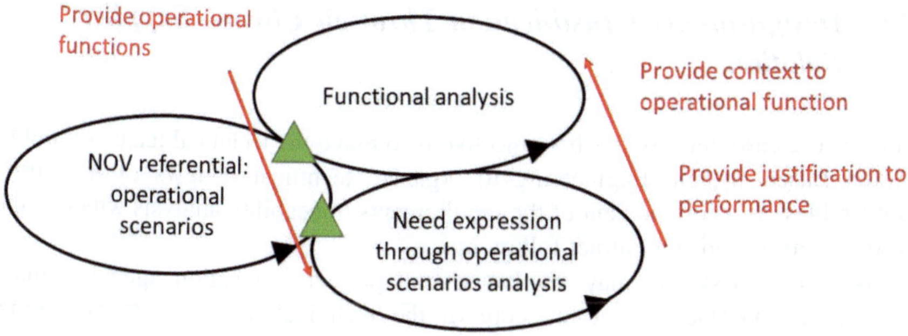

Fig. 9 Synergy between functional analysis and operational scenario analysis through our approach

4.2 *Studying Dependability—Functional Performances Trade-Offs*

This diagram suggests that effectiveness of this workflow would be widely bolstered whether dependability evaluation and performance evaluation trees be expressed from the same architectural properties stored into the NSV/NTV referential. The data specific to one of the two domain would then be store as attributes of the design's components. Another insight would be to evaluate the design on the same scenarios, e.g. for each scenario, evaluating the performance expected from the design to fulfill operational effectiveness objective, and also the dependability of the design under the conditions and threats of the same scenario, in order to trade-off dependability and performances at the scenario level, then aggregating these trade-offs on sets of scenarios (Fig. 10).

Fig. 10 Workflow to identify dependability/performance trade-offs

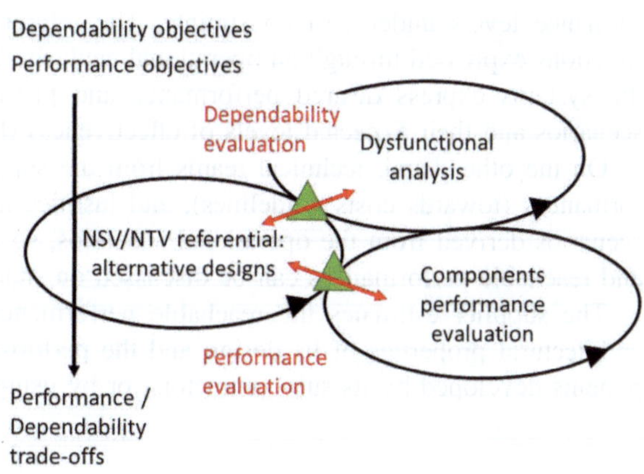

4.3 Design-to-Need Justification Through Client—Supplier Relation

For this use case, this approach's objective is to make the technical team commit to Model Based System Engineering to organize technical reviews of industry's deliverables, as a complement of the requirements traceability analysis which is the base for any technical contract follow-up.

As system-of-system may involve several project acquisition agencies, management of interfaces is at the core of the technical team's concerns. Some retro-engineering from the specifications would prove sufficient to fuel the NAF-based referential with the key representative operational scenarios involving all the parts of the system-of-systems of interest, and with the key architecture resources (mapped to operational actors). Then information exchange needs would be identified, then mapped to main communications channels.

A second step would be to redefine what is relevant into the supplier deliverables, so that the technical team would be able to evaluate whether the refined design proposed by the supplier is able to exchange the needed information through the already existing communication channels, and if not, how these communication channels could be improved. The main challenge to this second step is to find a quick way to select then import data relevant to the case from contract-driven deliverables.

4.4 Cost-Constrained Trade-off Analysis Through Client—Supplier Dialog

The purpose of this workflow is to build a referential to structure dialog on performance levels under cost constraints. The referential is based on operational functions expressed through an operational analysis. Then clients and end users of the systems express desired performance and justify them through operational scenarios and their expected levels of effectiveness (Fig. 11).

On the other hand, technical teams from the supplier proposes reachable performances (towards costs guidelines), and justifies them on technical-operational scenarios derived from the operational scenarios, so that the gap between desired and reachable performances can be discussed on shared scenarios.

The supplier estimates the reachable performance of the scenarios from the architectural properties of its design and the performance levels of the key components developed by its sub-contractors, or by using simulations.

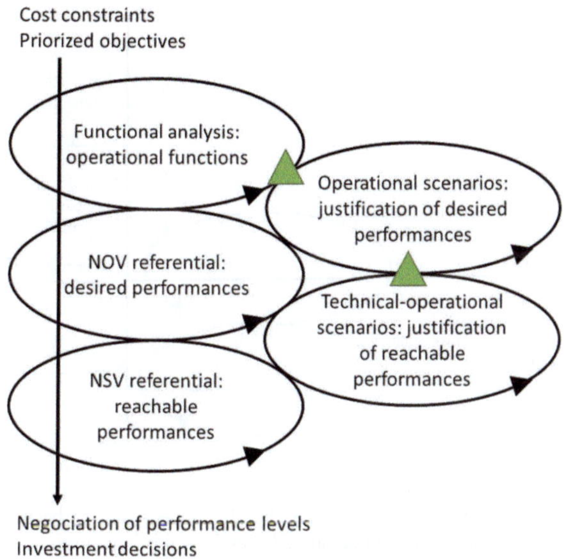

Fig. 11 Workflow to negotiate performance levels under cost constraints

4.5 Collecting Design Guidelines Through Explorative Engineering

Our approach highlights the redundancy between the need expression in the CONOPS based on representative scenarios derived from high-level operational scenarios, and the executable scenarios suitable for performance evaluation. So that the technical team leader gets high leverage to commit its team and its stakeholder to express the operational concept as a set of operational scenarios modeled through executable graphical notation, along with the set of parameters needed to simulate—and their variation ranges. Our approach also bolsters the need to capitalize as soon as possible key technical characteristics into the referential from domain-specific performance & technological studies, because this data is needed for evaluating performance through simulation (Fig. 12).

5 Case Study

This case study intends to illustrate the interest of our approach for the formalization of the decision process of a system-of-system or complex system acquisition project led by the contracting agency.

Our case concerns the Early Engineering phases of an acquisition project: the objective of this phase is to collect key elements of decision to define design orientations consistent with performance and cost objectives.

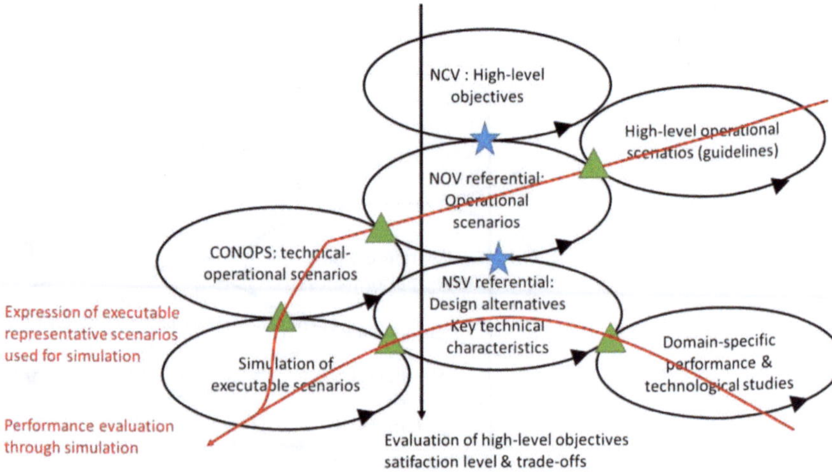

Fig. 12 Workflow suggesting the high added-value of basing the CONOPS on the expression of executable technical-operational scenarios for simulation

This collect is to be contracted through a study contract; this contract shall encompass the whole perimeter of the defense system and various issues (global performances for reference scenarios, safety, and definition of a system of support concept).

Because this contract encompasses the whole perimeter of the to-be defense system, and not only separate issues, it is an opportunity to define how our approach could structure the relation between the decision-making process of the acquisition project team of the contracting agency, and the System Engineering process of the development team of the contractor.

First step of the approach led to identify key NAF concepts needed for the study: concepts for operational need expression, concepts for architectural design, concepts for costing, safety concepts specific to the considered field, plus decision-making related concept that shall help keeping trace of the results of the decision-making process. Then it led to identify the interoperability formats needed to exchange data between the contracting agency tools and the contractor tools; the need of a State-of-the-art common graphical notation has been contractualized.

Second step of the approach led to identify the key consistency workflows expected from the contractor: cover of the operational need by the alternative designs, consistency of the design to the contracting agency decisions, identification of isolated objects into the referential. Impact analyses have been described that shall allow the contractor to identify synergies in order to anticipate the redundancies, contradictions and trade-offs between engineering domains.

Third step of the approach led to define viewpoints based on both a requirements referential and a NAF-based technical-oriented referential; these viewpoints address a specific preoccupations of the contracting agency. These viewpoints will be directly used to prepare the trade-off dossiers presented to the decision-makers, thus directly linking the System Engineering deliverables to the decision-making process.

Fourth step of the approach led to ask the contractor to describe its architecture evaluation methods and how these methods would be applied on the referential to propagate performances up to operational measure of effectiveness. Our objective is to get a high-level vision of how the performances and cost are affected by minor design alternatives.

6 Conclusion

This case study illustrated how our approach can serve as a conceptual framework to link Early Engineering activities to their purpose: enlightening the early decision-making process which set the cost-performance-delay-risk balance for the whole life-cycle of the system.

The use cases illustrated how our pragmatic approach contributes to prevent the technical team from investing too much time in the arcana of architecture modelling they are not familiar with; and instead, make them progressively provide data to the common referential; data, despite specific to their domains, expressed with team-approved, near-to-NAF concepts and graphical representations. Then, the collected data is imported by routines in the referential, and processed by consistency checkers. Examination of the data shall lead to identify synergies, conversing with engineers shall lead to identify relevant deliverables, and which missing data shall be produced into the referential to produce these deliverables.

To boast the effectiveness of this approach, a larger-scale System Engineering change management operation should focus on providing standard concepts mapping, automated tools to transfer and compare data between specific engineering domains and architecture modelling. Generic performance evaluation models or frameworks should be developed, that are able to evaluate through architecture properties most common performance and quality requirements.

References

1. Ernadote, D.: NATO IST-115. An automated objective-driven approach to drive the usage of the NAF framework
2. Luzeaux, D., Ruault, J.-R.: Hermes Science Publications Lavoisier. Ingénierie des systèmes de systèmes: Méthodes et outils
3. NATO Architecture Framework v3.0
4. NATO Architecture Framework v3.1 chapter 5

System Engineering Applied on Electric Power System for PHEV Applications

Benoît Beaurain, Ahmid El Hamdani and Joël Adounkpé

Abstract This article deals with the systems engineering approach applied to the Electric Power System (EPS) of the vehicle. We define how to characterize a system and how to describe the system following an analysis framework. This framework is applied to the EPS for PHEV application to give some systemic elements throw the Operational, Functional and Logical view. Despite some difficulties in the concrete application, the new paradigm brings benefits such as quality, complexity management and to improvements in the solutions' efficiency.

1 Introduction

In a highly competitive market, PSA PEUGEOT CITROEN strives for remaining at the front-edge of the vehicle usage experience. The obligation of operational efficiency and quality led the R&D Direction to activate the system engineering lever to master the growing complexity of the vehicles produced and to provide our customers with a relevant driving experience without compromise on reliability, availability or security. Following this systems engineering dynamic, the vehicle system design structure has been modified and exhibits three levels:

- Vehicle Design level.
- System Design level (engine, gearbox, electric power, steering system,...).
- Module/Component Design level (Sofware and Equipments).

B. Beaurain (✉) · A.E. Hamdani · J. Adounkpé
PSA Peugeot Citroen, 18 Rue Des Fauvelles, La Garenne Colombes 92250, France
e-mail: benoit.beaurain@mpsa.com

A.E. Hamdani
e-mail: ahmid.elhamdani@mpsa.com

J. Adounkpé
e-mail: ifedejoel.adounkpe@mpsa.com

G. Auvray et al. (eds.), *Complex Systems Design & Management*,
DOI 10.1007/978-3-319-26109-6_17

The development principles are roughly as follows:

- The "vehicle" level identifies, instantiates and allocates the functions allowing to ensure the vehicle benefits to the systems.
- The "system" level establishes the system solutions (system architecture) that meet the needs of the vehicle and allocates system level requirements to "components".
- The "component" level meets the system needs while ensuring the standardization of the solutions.

Redefining the "system" level should enable the group to:

- reduce the development costs;
- reduce the technical diversity;
- reduce the "time to market";

by articulating its best "module" strategy and the "system" strategies materialized by product or platform policies.

In the subsequent sections, we will present the underlying fundamentals which support this "systems engineering transformation" and the Electric Power System. Then we will elaborate on some systemic elements of the EPS according to the adopted systemic analysis framework. Finally, some practical lessons learned will be given.

2 System and Systemic Analysis Grid

2.1 Notion of System

Several definitions of the term "system" coexist. In the remaining of this article, we will use the (recursive) definition provided by Faisandier in [1] and illustrated by Fig. 1: "*A system is a collection of components such as people, hardware, software, materials, procedures or services, that are gathered and synchronised, so that their mutual interactions, using resources in a given environment, satisfies the needs and expectations, that are derived from its mission and objectives, themselves derived from its purpose*".

A system is first characterized by a purpose, a mission and objectives. These synthetic elements are detailed in terms of requirements and expectations which are refined in technical requirements. The mission is refined and described in the form of operational scenarios. The operational scenarios request exchange of material, energy and/or information between the system of interest and its environment (external systems). These exchanges allow to identify interfaces and interactions, and consequently the functional and physical boundaries of the system. The operational scenarios are carried out by functions grouped in a logical architecture. Functions are carried out by concrete components which compose a physical

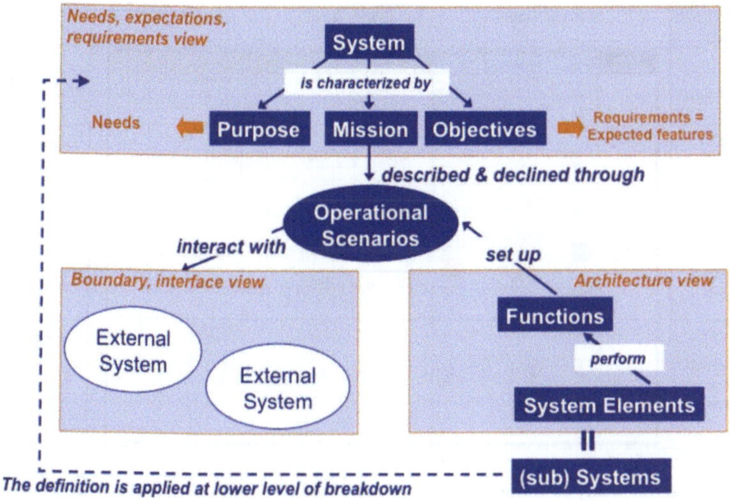

Fig. 1 System: a model of definition (from [1])

architecture. When the number of concrete components is important (more than ten), it is necessary to form subsystems. Each subsystem composing the studied system is a system in its own right, and thus is characterized by the same generic elements as the system of interest. This leads to a hierarchical composition of systems.

From this definition, we deduce that to engineer a system, one has to define a set of elements related to several views:

- The needs and requirements view: purpose, mission, objectives and operational scenarios;
- The architecture view: the logical architecture and the physical architecture;
- The boundary and interface view: the physical interface and the interactions with the environment;
- The system breakdown view.

2.2 *Systemic Analysis Grid*

Defining the previous elements requires a rigourous method. In [2], Krob describes a systemic analysis framework which allows to cover exhaustively all the systemic studies required to design a system.

The decomposition emphasizes three architectural views (easily mapped to the views defined by Faisandier), from the need to the solution through operational,

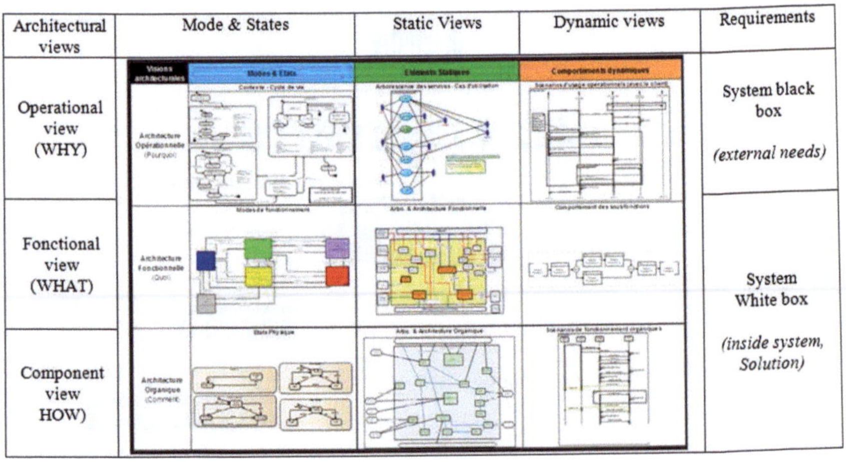

Fig. 2 "9 Views" analysis framework

functional and logical analyses. For each architectural view, the **modes and states**, the **static view** and the **dynamic view** of the system are described as illustrated in Fig. 2.

In the remaining of this paper, we will apply this systemic analysis framework to one of the systems which compose a vehicle: the Electric Power System.

3 Electric Power System (EPS) Overview

3.1 EPS Environment

The EPS exchanges information and energies with the gearbox, engine chassis, thermal systems and all system consumer of electrical energy (Fig. 3).

3.2 EPS Purpose, Missions and Objectives

The purpose of the EPS is to ensure the electric drive and the electrical autonomy of the vehicle. In order to guarantee this purpose the EPS must accomplish the missions below with the main associated objectives:

- Ensuring electric vehicle supplying
  - Energetic autonomy
  - Power levels (nominal, maximum…)
  - Grid voltage levels

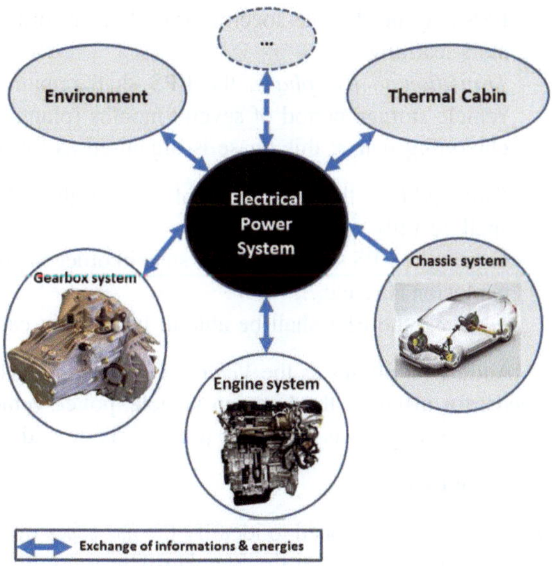

Fig. 3 EPS environment

- Ensuring Thermal Engine system electric driving
  - Engine Starting (start duration…)
  - Availability of engine starting
- Ensuring electric drive of the vehicle
  - E-drive autonomy
  - Power levels in e-drive mode (Acceleration and braking)

4 Systemic Analysis of the EPS

4.1 Operational View

(a) EPS Life cycle

The EPS life cycle is defined according to the vehicle life cycle from the design phase to the end of life.

The main phases to be considered in the EPS case are:

- *Manufacturing phase*: the EPS shall have minimum functionalities (for instance to ensure the first engine starting…) to guarantee the vehicle assembly and the

EPS may need some specific procedures in order to execute sensors and actuators learning.

- *Transition to use phase*: the EPS shall remain fully functional even with a vehicle storage period of several months (plant, showroom,…).
- *Operating phase:* this phase is organised as follows

  Standard Use: the main phase of using of the EPS by the final clients. The EPS shall be fully operational.
  Crash: the EPS shall be in safe state in order to prevent the risk of electrocution, explosion, fire etc.…
  Services: the EPS shall be able to be fixed in case of fault.

Additional phases as the *Vehicle Converting* phase can be analyzed for specific needs, for instance, the Fireman vehicle, police vehicle etc…

A state diagram is generally used to define and illustrate the system life cycle.

(b) **Use cases**

For each phase, we need to identify the use cases of the EPS. In this section we will show some examples in the Standard Use according to the EPS missions.

Figure 4 shows the main functionalities of the EPS in the standard use, each use case needs one or more of these functionalities. For instance, in a "1000 m Standing Start" use case, the vehicle objective is to realize a distance of 1000 m in less than a specified time, in this scenario the EPS will contribute to vehicle objectives by:

- Ensuring the Electric Supplying of the vehicle consumers with the electrical power need associated to this use case.
- Ensuring the Electric Traction with the mechanical power and energy necessary for this use case.

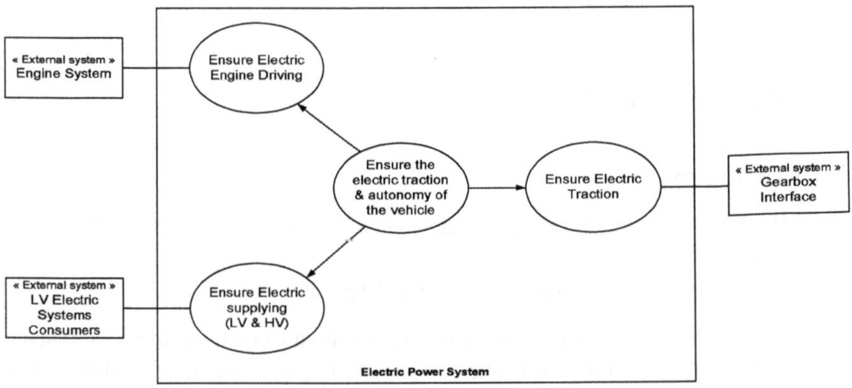

Fig. 4 Main functionalities of the EPS

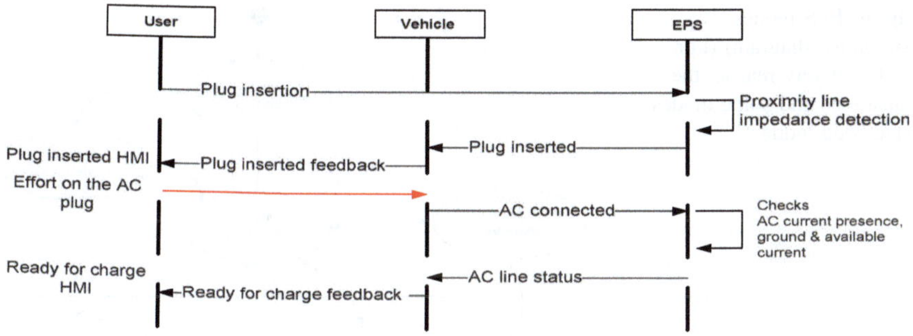

Fig. 5 Sequence diagram plug-in example (start of the execution)

In this use case, the EPS has not the possibility to receive energy from the Engine system or from the vehicle kinetic energy. The EPS shall execute the use case with its stored energy. A list of sizing use cases is defined in order to design the EPS.

(c) **Dynamic description**

To complete the use case specification, we need to describe its dynamic execution in order to identify the main interactions and interfaces between the EPS and the other systems of the vehicle. This description will also give the synchronization and time constraints. Figure 5 shows an example of Sequence Diagram for a Plug-in use case.

4.2 Functional View

(a) **EPS modes**

The analysis of all the use case allows us to identify and define modes. Each mode will impact the functional behaviour of the system. For instance, we have identified in the previous section, two use cases:

- 1000 m Standing Start: the EPS ensures the electrical traction and the electrical power supplying.
- Plug-in: the EPS ensures the electrical power supplying and recovers its nominal electrical capacities.

We can directly conclude that the EPS will have two functional behaviours in those two use cases. Consequently, we can define two specific modes:

- Electric Drive mode: The EPS is able to supply electrical energy to the vehicle equipments and is able to supply mechanical energy to perform the electric drive.

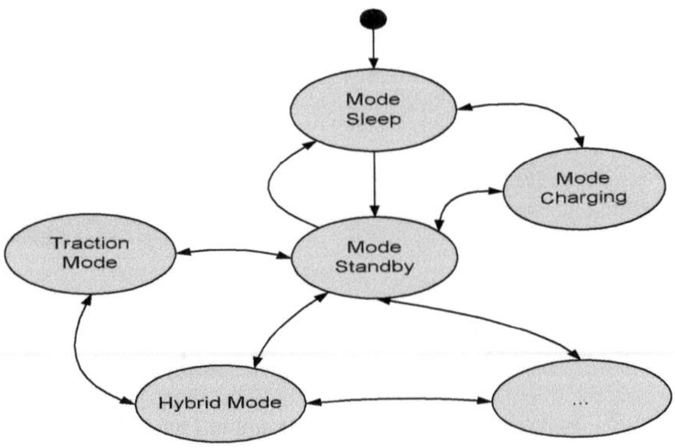

Fig. 6 EPS modes (simplified diagram) (For confidentiality reason, the transitions and some modes have been hidden)

- Charging mode: The EPS is able to supply electrical energy to the vehicle equipments and has the opportunity to recover its nominal capacities.

Finally, each identified use case has to be covered by a mode. Each mode can be considered as a behaviour sort of the use case. Figure 6 shows a simplified state diagram of the EPS modes in the standard use phase.

(b) **Functional Decomposition and Functional Architecture**

The first functional level is defined at the same time of the mode during the analysis of the use cases. We can identify what has to be done during the use cases execution and to identify the system functions. For the EPS we have three main functions:

- Supply Electrical Energy
- Supply Mechanical Energy for electrical drive
- Supply Mechanical Energy to the Engine System

The supplying can be positive or negative for acceleration or braking.

On the Requirement Development side, we have to describe the capacities of these three functions in each mode. The following matrix shows the functions/modes mapping (Fig. 7).

The next steps are the functional decomposition and the factorization in a functional architecture. For the decomposition, we use a simple pattern with colours. A priori, each function can be decomposed with "Transformation function" systematically associated with its "Control function" (Fig. 8).

		Supply Electrical energy	Supply Mechanical Energy for electrical traction	Supply Mechanical Energy to the Engine System
Modes	Sleep	x		
	Charging	x		
	Traction	x	x	x
	Hybrid	x	x	x

Fig. 7 Functions/modes matrix

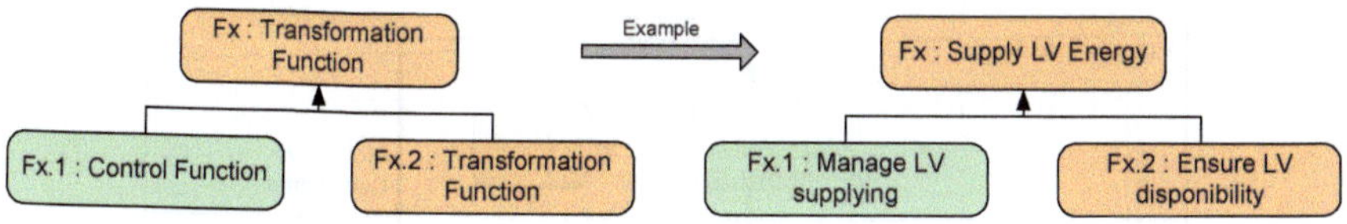

Fig. 8 Functional decomposition pattern

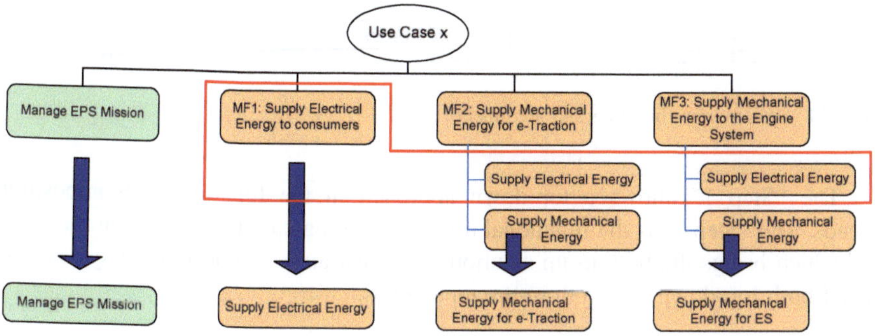

Fig. 9 1st decomposition level and factorization

This pattern allows us to build during the functional decomposition a hierarchized control of the system. The decomposition of each 1st level function may use similar "transformation function" due to the operational concept of the system. The 1st level EPS functions need electrical energy to meet the mission objectives. Figure 9 shows, the first level of decomposition of the EPS and the factorization.

We can analyse that each function use a similar function "Supply Electrical Energy" with specific needs. At this step, we have decided to factorize this function in the architecture. Figure 10 shows the 1st level architecture.

Then we continue to decompose until each function can be allocated to a logical component. Figure 11 shows the different steps of the functional architecture building and the alignment with the logical view.

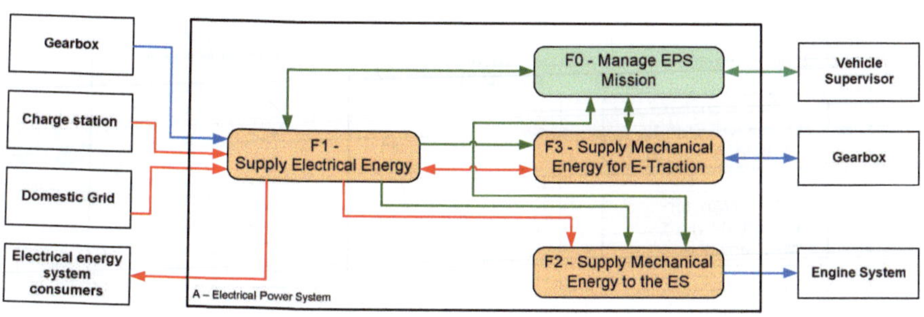

Fig. 10 1st level EPS architecture

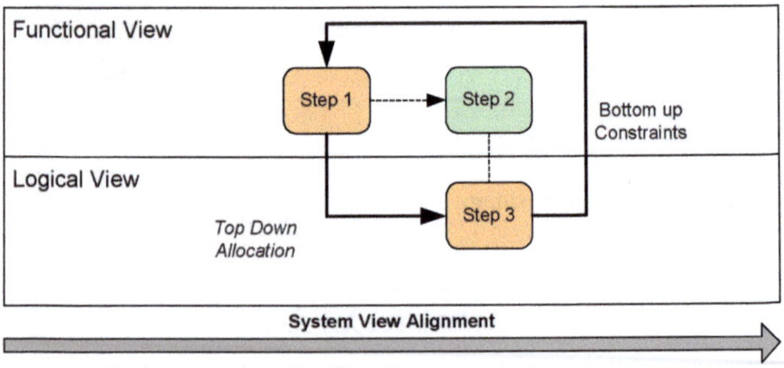

Fig. 11 System view building and alignment

The "Step 1" allows a top-down approach in the functional decomposition needs. The "Step 2" is the functional mirror of the logical view created in the "Step 3" which brings the bottom-up solution and constraints. To design the system, we need both top-down and bottom-up approaches.

4.3 *Logical View*

(a) **Functional allocation and Components Specification**

At this stage, we get a functional architecture ready to be allocated to logical components. Here the question is how to specify a logical component from a list of allocated functions without forgetting the basic rule *The requirements shall not express solution, excepted design constraint.*

Imagine functions allocated to one component. The risk is to specify each function independently and define unnecessary design requirements.. Our approach is to address the mission of the component, not its functions.. One way to succeed in this approach is to define from the allocated functions the modes of the component. Figure 12 shows a simplified example inspired of the EPS architecture.

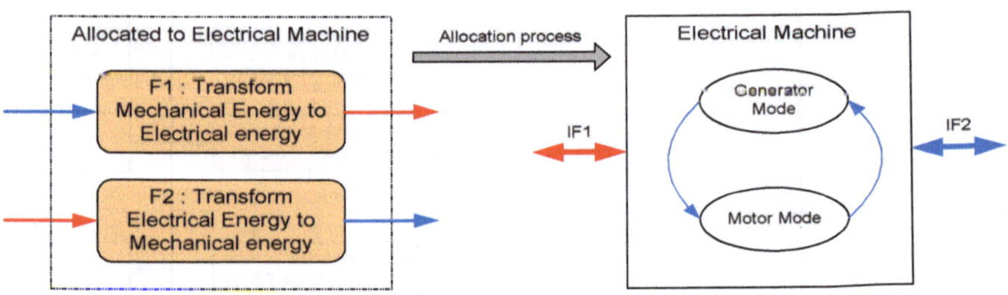

Fig. 12 Allocation process

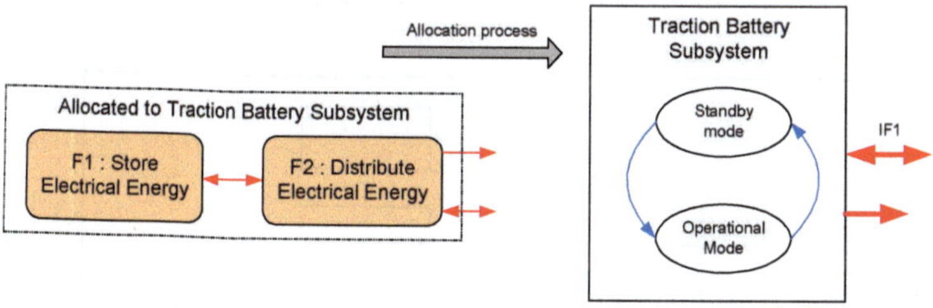

Fig. 13 Allocation process of functional chain

Both functions, transforming electrical energy into mechanical energy and transforming mechanical energy into electrical energy, are allocated to the same component *Electrical Machine* and we do not need to execute simultaneously the functions. Each function has interfaces and a characterized behaviour (power, efficiency, torque etc.…). It is a typical case, where the allocated functions in the functional architecture are the modes in the component specification. In this example the modes will be:

- Generator mode: to specify the component when it executes the function of transforming mechanical energy in electrical energy.
- Motor mode: to specify the component when it executes the function of transforming electrical energy in mechanical energy.

The functional interfaces are factorized in the same mechanical and electrical interfaces of the component. Another typical situation is the allocation of a functional chain to one component. As previously said, the risk is to specify in the component specification each function instead of specifying the need associated to the function chain as the whole (Fig. 13).

In this case, the modes are inherited from the modes of the allocated functions. In the previous example, "F2 Distribute Electrical energy" has two modes "on/off" and consequently the Traction Battery Subsystem has similar modes for functional chain as a whole. For complex subsystem, we generally have combinations of both typical cases previously explained.

(b) **EPS Logical architecture and Interfaces management**

The logical architecture describes the architecture of the component/subsystem of the system. Generally speaking, the EPS is composed of:

- Batteries subsystems to ensure the electrical energy storage
- DC/DC Electrical converters to ensure the adaptation of the voltages
- Electrical Machine subsystems to ensure the conversion Electric/Mechanic
- AC/DC Electrical converters to ensure the adaptation of the external voltage
- Electric/Electronic architectures to ensure the energy and information transport
- EPS Supervisor to ensure the control of the EPS components.

Figure 14 gives a simplified example of architecture of the EPS.

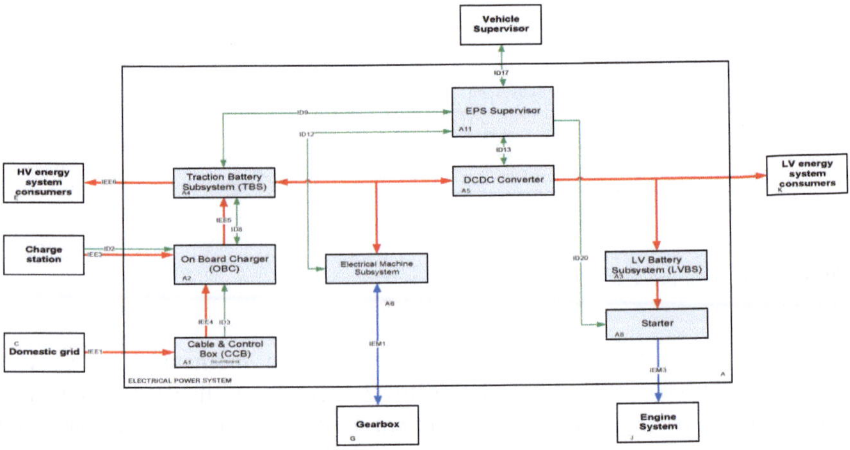

Fig. 14 Logical view of the EPS architecture

PUID - V	Requirement Text	SW Signal Name	Signal Range Min	Signal Range Max
HY4P-EPS-ICD-15 (0.9)	Couple MEL AV max nominal (EmTqLim_tqMaxNomFrntMch): Range: [0 2000]Nm Accuracy: 2Nm Initial Value: 0Nm	EmTqLim_tqMaxNomFrntMch	0	2000
HY4P-EPS-ICD-16 (0.9)	Couple MEL AR max nominal (EmTqLim_tqMaxNomReMch): Range: [0 2000]Nm	EmTqLim_tqMaxNomReMch	0	2000

Fig. 15 Interface control document EPS—DOORs

The logical architecture allows to identify and manage the interactions between the components of the system following the functional allocation. In this diagram, we identify external system, each component of the system and interfaces. In order to ensure the consistency, the interface definition of functional flow are managed in a specific document *Interface Control Document* in which we develop interfaces requirements for each exchanged data. The objective is to centralize the definition of the interfaces as a data dictionary. The traceability is ensured with the EPS Technical specification and the Components specification by using the DOORs links (Fig. 15).

5 Practical Lesson's Learned

The first application of the "9 views" framework needs an important effort to materialize the complete description of the system and in particular the dynamic description. Another difficulty that has been met is the definition of the boundaries of the systems of the vehicle design decomposition and its alignment with the organization. The SysMl modeling tools are not so much user-friendly and still needs to be improved to be efficient.

However, the systems engineering approach brought the capacity to grow the set of considered solutions, the capacity to manage the system complexity in terms of internal and external interactions and finally the quality of the final work products has been improved.

6 Conclusion

This article has provided an overview of the "9 views" framework applied on the EPS system. Despite some difficulties in the application, the method brings benefits and is the prerequisite for the next steps to grow the design efficiency which are:

- Defining the EPS product line to organize the reuse of system design elements and system studies;
- Developing the Model Based System Engineering, consistent with the product line, and **articulated with** the current system analysis supported by the modelling & simulation.

This will actively contribute to the success of the transforming of the Research and Development department, necessary condition to achieve the objectives of the group PSA Peugeot Citroen.

References

1. Faisandier, A.: Notions de Système et d'Ingénierie Système. Eng. Architecting Multi. Syst. **1** (2014)
2. Krob, A.: Eléments de systémique—Architecture des systèmes
3. AFIS.: Bonnes Pratiques en Ingénierie des Exigences. Collection AFIS. Editions Cépaduès (2012)
4. AFIS.: Découvrir et Comprendre l'Ingénierie Système. Collection AFIS. Editions Cépaduès (2012)

Operational Analysis of Virtual IP Multimedia Subsystem (IMS) Through a Model-Based Architectural Framework

Arevik Gevorgyan and Peter Spencer

Abstract Telecom/IT convergence is transforming network architectures and cost structures. Lack of methodological support within an equipment provider organization hinders heavily the possibility of efficient transitions from traditional monolithic to virtual architectures and their market insertion. For any evolutions, especially innovation, operational analyses have a vital importance. We propose to base our approach for a systematic operational analysis upon the following metaphors: (1) an adapted Architectural Framework (in our case, inspired by SAGACE), incorporated with (2) PESTEL (environmental) and subsequently FURPSE (software characteristics) analyses frames, (3) leveraging on Model-Based Systems Engineering. A common consistent language and format for structuring and relating system's operational, functional and physical views allow handling in a holistic and integrated manner evolutions and complexities of the system and its environment throughout decisions spectrums and levels. We case study virtual IP Multimedia Subsystem (essential for communication services across networks). Identified operational invariants are inputs of critical importance for iterative functional and decisions trade-off analyses in accordance to market, technological and other perspectives.

Keywords Network functions virtualization · IP multimedia subsystem · Operational analysis · Complex systems architecture · Architectural framework · Model-Based systems engineering

---

A. Gevorgyan (✉) · P. Spencer
Laboratoire d'Informatique (LIX), École Polytechnique, Route de Saclay,
Palaiseau 91128 France
e-mail: gevorgyan@lix.polytechnique.fr; arevik.gevorgyan@alcatel-lucent.com

P. Spencer
e-mail: peter.spencer@alcatel-lucent.com

A. Gevorgyan · P. Spencer
Strategy—IP Platforms, Alcatel-Lucent International, 148/152 Route de La Reine,
Boulogne-Billancourt 92100 France

G. Auvray et al. (eds.), *Complex Systems Design & Management*,
DOI 10.1007/978-3-319-26109-6_18

1 Introduction

Network Functions Virtualization (NFV) is the greatest transformation in Telecommunication industry and, in particular, for Equipment Providers. An assumption made by both fixed and mobile operators is that virtualizing networks will fundamentally change architectures and cost equations of how networks are built and managed.

Our study belongs to Alcatel-Lucent's current initiatives to: *(a) understand the impacts of transformation in technological and business terms; (b) manage the transition from traditional monolithic to (optimal) virtual architectures; (c) and improve its systems development practice to achieve efficiency.*

The most complex and important NFV instance for Alcatel-Lucent is the IP Multimedia Subsystem (IMS), essential for communication services across networks and the only for Voice over 4G (further 5G).

Correlated with its development practices, any evolution of the system seeks to maximize its success factors (where enterprise strategy and technological capabilities (the "know-how") play a key role), minimize risks and optimize CQFD (cost, quality, functionality, delay) parameters. Achievement of an optimal virtual IMS architecture is difficult in light of the complexities inherent to the system and its evolving systemic environment, especially with the absence of an adapted unified multi-disciplinary approach.

To define an architecture best adapted to the context, it is needed to analyze in details and satisfy a network of constraints of a very different nature (i.e. environmental, system, organizational etc.).

Operational analysis is vitally important and iterative in this journey. It defines in a non-ambiguous way virtual IMS external interfaces, mission, stakeholders, their needs, contexts, uses cases, scenarios, etc.

Traditional Systems Engineering (SE) methods (also non-standard to a specific domain), opting for per physical-unit decompositions, design systems that lack knowledge upon their environments, subsystems, interrelations of views, etc. Moreover, Systems Integrators, concerned by governance issues, usually prefer to decompose around managerial criteria. Such practices generate a significant gap between the problem and the solution space, triggering a bucket of different kind of issues. Evidently, they are no longer efficient or reasonable in light of the NFV innovation.

We explore examples of complex industrial projects best practices and advancements in modern Systems Engineering. Hence, through probes, we propose a set of incorporated metaphors to better underpin the multi-facet aspects of the problem and to compile an integrated holistic analysis approach.

We constitute our approach upon: *(1) an adapted Architectural Framework (inspired by SAGACE), incorporated with (2) environmental (PESTEL) and subsequently (software) system characteristics (FURPSE) analysis frames, (3) leveraging on Model - Based Systems Engineering (MBSE)* to model the system views (operational, functional, physical), its overall environment and direct market specifics in accordance to desired system properties. This practice also aims to improve cross-functional collaborations, focusing on evolutions-driven perspectives.

To structure our approach: in Sect. 2 we briefly discuss the virtual IMS system, its context and inherent complexities. At this step, we also define the initial measures for our further analyses. This basis motivates our interest in the proposed method. In Sect. 3 we present the operational analysis procedure and its outcomes. We demonstrate it through a brief example from the IMS "Operate" lifecycle phase, also to better highlight the NFV innovation features. In Sect. 4 we explain the importance of operational analyses outcomes for functional and decision making perspectives from methodological and practical standpoints. We also discuss our vision upon strategic and research prospects.

2 Preliminary Background

Design, development and integration of virtual IMS is difficult due to its context and inherent complexities. This brief contextual overview, instantiated from PESTEL (Political, Economic, Social, Technological, Environmental and Legal) layers, including few more indispensable axes (Regulatory, Competition, Organizational) shall help to justify the choice of our method, also to deduce initial Measures of Effectiveness (MOE) for our analyses and estimations.

2.1 Virtual IMS Brief Context: Axes of Complexities

IP Multimedia (Core) Subsystem is a network architectural framework for delivering multimedia communication services across any types of networks. An IMS solution is made of multiple network components, which are the Network Functions: as P/I/S-CSCF, MGC, CTS, HSS, CCF, MRF, BGCF, Application Servers, etc., that can be delivered by different Telecom Equipment Providers or organizations within one Telecom Provider. This used to result in a solution made of a multitude of different hardware boxes, where even though the hardware used may actually be the same, it could not be shared by the IMS components unless they used a specific ad hoc middleware. Moreover, although the IMS components are actually pieces of software, Operators could not purchase the licenses of the IMS components and run them on standard and common hardware (that the Operator would own). The initial purpose of IMS Cloudification and Virtualization effort has been to resolve the few major issues, as:

1. *Propose to Customers an IMS solution that is compact and easy to deploy and can run on a monolithic and homogeneous hardware that is used optimally*

2. *When the hardware is provided by ALU along with the IMS solution, allow any 3rd party software to run on this hardware as on a Cloud*
3. *Allow Customers to purchase "software only" IMS components, or even an entire "software base" IMS solution and run it on their own data center, or on any appropriate 3rd party Cloud*
4. *Further on, to significantly maximize and optimize network capacities (through virtual machines deployments and orchestration) and support new services (new network functions deployments)*

Competition

Telecom players face a fierce competition from Internet services and Over-the-Top (OTT) content providers, Google, Netflix, Amazon, large groups with high performance international infrastructures all over the world. Possessing advanced architectures and being to some extent less restricted by regulations, they are capable to offer equivalent communication services in a more efficient way. Telecom operators are forced to create differentiating offers with better, cost effective services, simultaneously reconsidering their architectural choices to be capable to deliver new features and handle massive traffic rates [1].

Market

When we reflect upon the development of Cloud communications, collaborations around it, everything connected, all within a frame of "intelligent" contextual communications, we realize that there is a range of ecosystems that need to be considered, supported or created. The complexity, at first, results from the number of stakeholders and heterogeneity of their environments. For instance, in Telecom Operator organizations the accumulated legacy or different levels of technological maturity cannot be easily transformed or managed [2].

Regulatory

Standardization bodies play a significant role in the Telecommunications industry. They guarantee the interoperability between vendors and create a common ground, where all players can push forward ideas to direct the industry.

Alcatel-Lucent follows the 3GPP's IMS [3] and ETSI ISG's NFV standards [4]. The problem question for Alcatel-Lucent is how to organize and deliver the virtual network functions (VNFs) efficiently.

Technological

The transition to the "full NFV" solution has still a considerable path towards its implementation. Infrastructure/Platform as Service (IaaS/PaaS) architectures are just starting to mature in the IT space, but are slower to settle in the Telecom space. Moreover, network function specificities create an additional set of constraints, as compared to traditional IT virtualization.

Though all profit from the cloud concepts on COTS hardware, the real-time communications networks and their supporting infrastructures are still being studied by both vendors and standardization bodies. Therefore, while specifying the NFV architectures in the RFP documents, customers are still in their experimentation cycle when it comes to the production environments.

A real difficulty is to ensure that everything works together: i.e. multi-vendor platform and different deployment scenarios (end-to-end infrastructure or software integration only). The scale and complexity of telecom networks requires a level of commonality beyond any single organization or entity.

Economic & Legal
As initial estimations indicate, Cloud helps to save the principal portion of CAPEX, though OPEX (pre-/post-integration services expenses) remains very high. Ideally, the time spent at a customer site shall be reduced from years - months to few days-hours. Risk management and contractual constraints is one key parameter restraining any changes in the TCO.

Organizational
Design and developments take place within the historic silo units, respectively mirroring their "old" traditional architectural choices as per physical unit compositions. Knowledge and information sharing within the project is a struggle.

Evidently, such practices hinder the opportunity to better benefit from Cloud capabilities.

To conclude upon the context at this point, we may already deduce some principal Measures of Effectiveness (MOE) for our analysis: (1) Maximize Network Capacity, (2) Minimize Expenses, (3) Maximize System Autonomy. More MOEs could be defined in line with the objectives of the study.

2.2 Motivation of Our Approach

IMS is a complex system, which operates in an evolving complex systemic environment. Numerous stakeholders and systems with different, also antagonistic "requirements" and lifecycles constitute the IMS environment. Any evolutions/impacts may be viewed from system of systems (SoS) perspective. One of the most important and difficult aspects in System Engineering is the specification and management of interfaces. It is critical to define clearly and correctly the borders between the system of interest and external systems with which it interacts, in order to avoid the impacts of their evolutions. Vital is to analyze correctly the system views and interrelations. Evidently, any mistakes in the analysis process lead to a solution, integration of which may be difficult or impossible.

Throughout our study we explore the related state of the-art and examples of complex systems projects best practices as a comparative basis. For instance, [5–7], etc. demonstrate the application and benefits of an Architectural Framework and MBSE applied in automotive, aerospace, etc. industries. We summarize on theoretical ground in Sects. 2.2.1 and 2.2.2.

2.2.1 Why an Architectural Framework?

Besides structuring and relating views, architectural framework improves the collaboration practice in complex systems projects. It ensures completeness, traceability, re-use and justification of top-down and bottom-up decisions throughout the system lifecycle.

The Architectural Framework concept was first proposed by [8]. Few prominent examples immerged later are DoDAF, developed by the US Defense department and MoDAF [9], developed by the British Ministry of Defense [10]. Other frameworks also exist: Domain Mapping Matrix, Design Structure Matrix, Quality Functional Deployment/House of Quality, Unified Program Planning, Axiomatic Design, CLIOS (Complex, Large-scale, Interconnected, Open, Socio-technical System), which are discussed in details by [3] and which limitations are explained. These frameworks do not capture the domain, technical, social, time concepts, as well as their interactions.

In our study we choose to refer to an adapted SAGACE framework, originally proposed by Penalva [11]. It constitutes the main principles for an iterative and incremental application for a complete design:

- *Modeling approach*
- *Graphical modeling language*
- *Matrix of nine points of view*: Operational, Functional and Structural, all three refined by three time perspectives. In our case, we refine the views by behavioral instead of time perspective, in order to use the SysML language of modeling, as explained in [12, 13].

2.2.2 Why Model-Based Based Systems Engineering?

As stated by the International Council on Systems Engineering (INCOSE) INCOSE "Systems Engineering Vision 2020 vision [INCOSE 2007], Model-Based Systems Engineering (MBSE) is one of the most prominent emerging practices in SE. MBSE aims to build integrated models with a holistic view to ensure the completeness of the design through retraceable requirements and knowledge across the project participants.

[10] illustrates the existing modeling languages appeared since the 1960s: ADL, AUTOSAR, UML, SysML, MARTE, EAST-ADL. [14] explains that the standardized modeling languages, as OMG SysML and UML are more effective for collaborations. We will use SysML™ described in details by [15, 16]. Implementation of models within the Architectural Framework is explained by [12, 13].

3 Operational Analysis

Important is to note, that introduction of any evolution, even of a feature size, disrupts the existing environment and may generate undesired consequences, if not previewed and addressed in advance. To develop and inject optimally the virtual IMS into the market, a broad range of specific needs and requirements must be considered and satisfied. To define an architecture best adapted to the context, it is needed to analyze in details a network of constraints that the system shall satisfy. These constraints have a very different nature and could be constituted within two conceptual network levels, which are to be incorporated with each other:

I. Those resulted from the Environment, in which the System and its Organization are immigrated. They are the factors of PESTEL (Politic, Economic, Social, Technological, Ecological, Legal), following the terminology of INCOSE.
II. Those specific to the (Client) Organization that uses the System and based on which it can operate. They are characterized by FURPSE (Functionality, Usability, Reliability, Performance, System Maintainability, and Evolution), following the ISO/IEC 9126 norm.

It is useful to follow the trajectories of corresponding processes [following SEBoK standards] throughout the system lifecycle, also to eliminate the constraints.

Operational analysis procedure leads to invariants identification: mission, external interfaces, stakeholders, needs, operational contexts, use cases, scenarios etc. Analysis starts from the system's environment modeling in order to identify the system's external interfaces: at this stage, stakeholders and their needs are identified. A refinement procedure is undertaken until precise elimination of associated micro needs. The system mission shall be clarified at this point. The second part of operational analysis is the operational analysis core: identification of operational contexts, use cases and scenarios.

3.1 Virtual IMS Environment Modeling

3.1.1 Identification of Stakeholders

Clear and correct definition of external interfaces is a crucial step before reaching the internal interfaces specification and their optimization phase. Multiple direct and indirect stakeholders (customers, end users, equipment and software manufacturers, IT companies, regulatory bodies, suppliers, financial institutions, etc.) are involved in the virtual IMS environment.

We are defining and categorizing the stakeholders and their inherent complexities within the PESTEL frame (Political, Economic, Social, Technological, Environmental and Legal) [INCOSE], including Regulatory, Competition and Organizational axes, as indispensable. When modeling the complete environment,

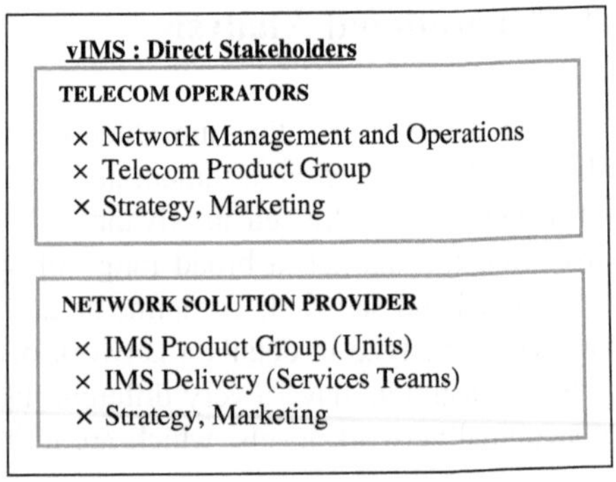

Fig. 1 vIMS direct environment

we would organize the stakeholders in seven-eight abstract classes (instantiated from PESTEL) based on their role importance and impact of each external system. However, to remain unambiguous, in this case study, given the number of external interfaces, also the significantly different dynamics of impacts, we focus only on two direct stakeholders (Telecom Operators and Network Solution Providers) (Fig. 1).

3.1.2 Analysis and Refinement of Needs

The broad range of stakeholders needs (according to market and technological criteria, as of primary importance compared to other dimensions and their development dynamics) are to be defined and refined iteratively throughout the system lifecycle phases. We characterize them in terms of desired software (functional) architecture properties, as FURPSE (Functionality, Usability, Reliability, Performance, System Maintainability, and Evolution) [ISO/IEC 9126 norm]. Needs are defined more precisely only in the late stage of the development phase, which is one of the major difficulties encountered in the analysis process. Below is presented the Table 1 with Macro Needs examples.

A Macro need refinement example is given below (Table 2).

Table 1 Macro needs definition

N1	Operators want a system that will support significantly higher traffic loads
N2	Operators want a robust system
N3	Operators want assurance for the maintenance and support
N4	Operators want capabilities to easily deploy/support new applications/services
N6	Operators want significant savings in CAPEX/OPEX
N7	Operators want operational easiness: i.e. to drastically reduce time to market
N8	Providers want to follow existing standards and regulations
…	………

Table 2 Macro need refinement

N2.1	Operators want an automatic adjustment of resources allocation for traffic growth and de-growth
N2.2	Operators want maximal availability and speed for huge traffic rates
N2.3	Operators do not want to feel the any limitations of the system
N2.4	Operators want predictable behavior of network functions
…	

Table 3 Requirements definition

Req.	Requirements derived from micro needs (ex. for N3)
R3.1	The SLAs have to be strictly respected
R3.2	The system has to be tested in accordance to specified standards
R3.3	Precise estimation of resources (CPU, memory) allocations
R3.4	Evaluation of inconsistencies for multi-vendor solution
….	

The refinement is done until reaching the level of clear, precise, measurable, and quantifiable needs, which are translated into requirements (Table 3).

3.2 *Operational Analysis Core*

Any system operates within certain operational **contexts**. The contextual diagram demonstrates the possible transitional cases for any situations to be handled by the system, associating them with the states of the interacting external systems. This way the contexts of external systems are to be identified as well.

External systems involved statically in a given context and their specific interactions with the system of interest are the **use cases**. The dynamics of interactions between external systems in a given context and our system of interest are the **scenarios**.

To describe the operational contexts we use *State Machine diagrams,* for use cases *Use Case diagrams,* and for scenarios *Sequence diagrams*, as explained in [13].

3.2.1 Operational Contexts

The study of the operational contexts shall retrace the system lifecycle. The lifecycle phases of Virtual IMS: Design and Development, Integration and Deployment, Maintenance, Operate/Use, and Disruption (which can happen in case of transition to a new technology) (Fig. 2).

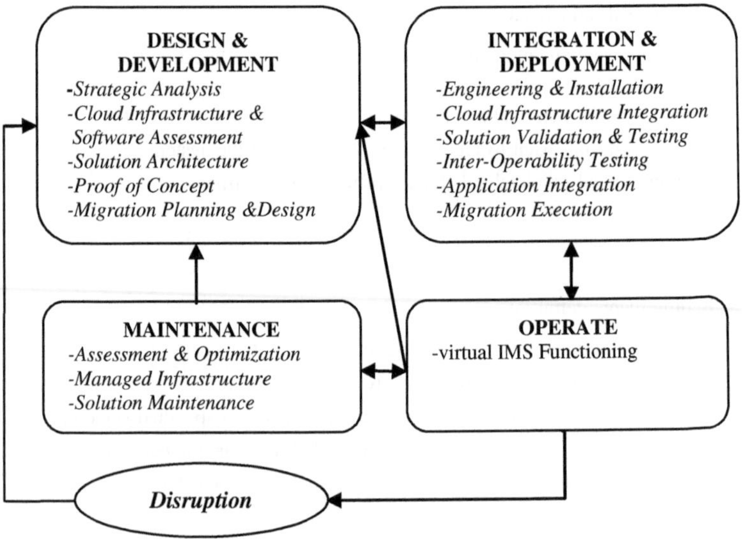

Fig. 2 vIMS lifecycle

In our example, instead of a complete lifecycle analysis, we focus on the states of external systems that directly affect the operability of our system. An example from “Operate” phase illustrates a case when the virtual IMS interacts with the supporting hardware infrastructure (Fig. 3).

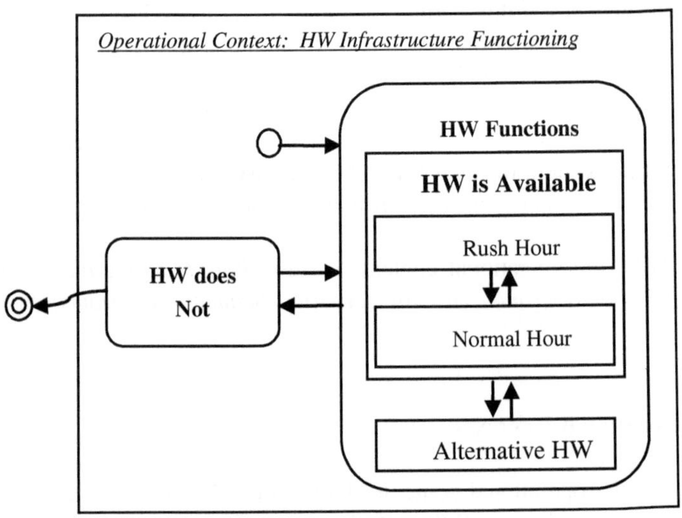

Fig. 3 ex. operational context

Table 4 ex. operational states: vIMS HW

<table>
<tr><th colspan="3">States of vIMS</th><th colspan="3">States of hardware infrastructure</th></tr>
<tr><td rowspan="5">OPERATE phase</td><td colspan="2">vIMS does not function</td><td colspan="3"></td></tr>
<tr><td colspan="2" rowspan="3">vIMS is deployed</td><td rowspan="2">HW functions</td><td>HW is available</td><td>Normal hour</td></tr>
<tr><td>Alternative HW</td><td>Rush hour</td></tr>
<tr><td>HW does not function</td><td colspan="2"></td></tr>
<tr><td>vIMS functions</td><td>New virtual machines Instantiated</td><td colspan="3"></td></tr>
</table>

3.2.2 Operational States

Combination of the different states of the virtual IMS and its supporting hardware infrastructure explains the possible situations needed to be taken into consideration, as the example below illustrates (Table 4).

3.2.3 Operational Scenarios

An example of a scenario in the "Operate" phase:

- the system lacks capacity (memory or CPU), as a result cannot handle the needed amount of traffic at requested speed
- new virtual machines are initiated (automatically) to gain the needed capacity
- the system transmits the needed amount of traffic with the new VMs

This case study shall not be considered as exhaustive. It is only illustrative and provides basis for consequent analyses procedures.

4 Results and Perspectives Discussion

Operational analysis invariants contribute to the functional analysis and related decisions tradeoffs. Based on the use cases and related scenarios analysis, the macro functions are first defined, serving as a basis for the Functional Breakdown Structure (FBS). Further on, based on scenarios analysis macro functions are refined into micro functions. Consequently, the FBS is refined and completed by behavioral and functional modes. In order to define a functional architecture best adapted to the context, stakeholders needs and satisfies optimally the desired constraints (i.e. as Cost, Quality, Functionality, Delay), we shall undertake tradeoffs according to technological and market choices, in prior. Numerous criteria and their interrelations are to be taken into consideration.

In our further studies, for functional choices derivation, assessment and optimization, we are interested to investigate decision making techniques, including decisions analysis processes that found application within SE practice. The objective is the solution for an optimal virtual IMS architecture. Simulations are envisaged for further evaluations.

5 Conclusion

In this paper we explained the strategic importance of the transformation initiated in the Telecommunications domain by Network Functions Virtualization. We undertook a case study on operational analysis of virtual IP Multimedia Subsystem. We explained the importance of a consistent adequate operational analysis for any evolutions, especially innovation. We proposed a method for a holistic integrated analysis based on complex systems architecture framework and model-based systems engineering technique, incorporated with PESTEL and FURPSE analyses models. Our method is inspired from examples of complex industrial projects best practices and advancements in SE practices. We highlighted the usefulness of the proposed approach for managing the complexity and evolutions of the system and its environment through a better cross-functional collaboration. We discuss perspectives for functional analysis and decisions trade-offs for an optimal virtual IMS architecture design.

References

1. Cisco. Evolution of the Mobile Network. White Paper (2010). http://www.cisco.com/c/en/us/solutions/collateral/service-provider/mobile-internet/white_paper_c11-624446.html
2. 4G Americas. Bringing Network Function Virtualization to LTE. White Paper, November 2014
3. http://www.3gpp.org/
4. ETSI GS NFV: Network functions virtualization (NFV), use cases specification 004 V1.1.1
5. Chalé Góngora, H.G., Gaudré, T., Tucci-Piergiovanni, S.: Towards an architectural design framework for automotive systems development. In: CSD&M (2013)
6. Berrebi, J., Krob, D.: How to use systems architecture to specify the operational perimeter of an innovative product line. INCOSE **22**(1), 84–99 (2012)
7. Doufene, A., Krob, D., Chalé Góngora, H.G., Dauron, A.: Model-based operational analysis for complex systems - a case study for electric vehicles. INCOSE **24**(1) (2014)
8. Zachman, J.A.: A framework for information systems architecture. IBM Syst. J. **26**(3), 276–292 (1987)
9. https://www.gov.uk/mod-architecture-framework
10. Björnander, S., Grunske, L.: Architecture Description Languages for Automotive Systems—a Literature Review. Technical report: C4-01 TR M49, 30 July 2008
11. Bartolomei, J.E., Hastings, D.E., de Neufville, R., Rhodes, D.H: Engineering systems multiple-domain matrix: an organizing framework for modeling large-scale complex systems.

MIT. Published online 10 October 2011 in Wiley Online Library (wileyonlinelibrary.com). Accepted 24 Feb 2011
12. Krob, D.: Eléments d'architecture des systèmes complexes. In: Gestion de la complexité et de l'information dans les grands systèmes critiques, pp. 179–207. CNRS Editions (2009)
13. Krob, D.: Enterprise Architecture, Modules 1–10, Ecole Polytechnique, 2009–2010 (personal communication)
14. Estefan, J.A.: Survey of Model-Based Systems Engineering (MBSE) Methodologies. Report of INCOSE MBSE Focus Group, Rev. A, 25 May 2007
15. Friedenthal, S., Moore, A., Steiner, R.: A Practical Guide to SysML—The Systems Modeling Language. Morgan Kaufmann, Burlington (2008)
16. Weilkiens, T.: Systems Engineering with SysML/UML—Modeling, Analysis, Design. Morgan Kaufmann Publishers (2008)
17. INCOSE. Systems Engineering Vision 2020 (2007)
18. Chalé Góngora, H.G., Dauron, A., Gaudré, T.: A commonsense-driven architecture framework. A car manufacturer's (naïve) take on MBSE. In: INCOSE (2012)
19. Honour, E.C.: Understanding the value of systems engineering. In: INCOSE (2004)
20. INCOSE. Systems engineering handbook. A guide for system lifecycle processes and activities. In: International Council on Systems Engineering (INCOSE), San Diego, CA, January 2010
21. PESTEL. PESTEL Analysis of the Macro-Environment. Oxford University Press, Oxford (2007)
22. Penalva, J.M.: La modélisation par les systèmes en situations complexes. PhD thesis, Université de Paris 11, Orsay, France (1997)
23. http://dodcio.defense.gov/Portals/0/Documents/DODAF/DoDAF_v2-02_web.pdf
24. Saaty, T.L, Vargas, L.G.: Models, methods, concepts & applications of the analytic hierarchy process. In: International Series in Operations Submitted for publication in the Journal of Systems Engineering, April 2011. Research and Management Science, Springer, Berlin (2000)

Urban Lifecycle Management: System Architecture Applied to the Conception and Monitoring of Smart Cities

Claude Rochet

Abstract At date, there is no standardized definition of what a smart city is, in spite many apply to propose a definition that fit with their offer, subsuming the whole of the city in one of its functions (smart grid, smart mobility…). Considering the smart cities as an ecosystem, that is to say a city that has systemic autopoeitic properties that are more than the sum of its parts, we develop an approach of modeling the smartness of the city. To understand how the city may behave as a sustainable ecosystem, we need a framework to design the interactions of the city subsystems. *First* we define a smart city as an ecosystem that is more than the sum of its parts, where sustainability is maintained through the interactions of urban functions. *Second*, we present a methodology to sustain the development over time of this ecosystem: Urban Lifecycle Management. *Third*, we define the tasks to be carried out by an integrator of the functions that constitute the smart city, we assume public administration has to play this role. *Fourth*, we present what should be a smart government for the smart city and the new capabilities to be developed.

Since the advent of the "death of distance" with the revolution of transportation by the middle of the 19th century, the appearance of networks of infrastructure technologies and the spread of the telegraph that transformed the government of the city, critical obstacles to the growth of cities were removed. Today digital technologies amplify this move, providing new tools such as smart phones that became a digital Swiss knife that allows inhabitants to be active actors in the city life, communicating and coordinating with each other, using and feeding databases. Doing this, digital technologies may produce the best and the worst. The point is each city contains the DNA of its own destruction. Smart cities digital infrastructure amplifies the possibilities of manifestation of discontent, worsening the gap between have and have-nots. Smart cities incur the risk to become the digital

C. Rochet (✉)
Professeur des universités, Aix Marseille Université, IMPGT AMU CERGAM EA 4225, Service de Coordination à l'intelligence économique—Ministère de l'Economie et des Finances, Paris, France
e-mail: Claude.rochet@univ-amu.fr

G. Auvray et al. (eds.), *Complex Systems Design & Management*,
DOI 10.1007/978-3-319-26109-6_19

analogue of the Panopticon Jeremy Bentham's prison design (Townsend 2013). Therefore, architecting the city as a living system is as well technical as political.

This paper is based on case studies carried out in various countries, analyzed though the lens of complex system architecture, to envisage how these competencies may be adapted to the modeling of smart cities.

1 Ancient Cities Were Smart

Far as back as 1613, the Napolitano Antonio Serra analyzed the city as the place where activities with the biggest increasing returns take place, with a strong correlation between economics and politics [17]. The frescoes of the Siena town hall by Ambroggio Lorenzetti depict "the good government" as a dynamic equilibrium between intense economic activities and an active political life that gives the people of citizens the power to rule the city according to the principles of the common good. Contemporary evolutionary economics correlates the evolution of institutions with that of economic activity (Reinert 2012). This evolutionary process was secured thank to learning feedback loops which duration was generations, the latest learning from the former to design the city in a way to optimize interactions between activities.

The growing complexity of cities and the predominance of top-down urban planning made us forgetful of these lessons from the past. In their analysis of present smart cities initiative, Neirotti (2013) notice that there is no practice that encompasses all the domains, hard and soft, of the cities. The most covered domains are hard ones: transportation and mobility, natural resources and energy. Government is the domain in which the cities report the lowest number of initiatives. More, in the present smart cities research program, there is an inverse correlation between investment in hard and soft domains, smart government being still the poor relative in smart cities initiatives and cities that have invested in hard domains are not necessarily more livable cities. In fact, two models emerge from Neirotti et al. survey: one focused on technology (with a strong impetus of technology vendors) and one focused on soft aspects, the hard model being dominant. The problem is there are no vendors for soft domains apart the citizens themselves whereas systemic integration relies on soft domains, mainly taking in account the context and valuing social capital.

1.1 What Is an Urban Ecosystem?

A smart city is more than the sum of "smarties" (smart grids, smart buildings, smart computing…) although it is referred to in the absence of a precise and operational definition of what a smart city is [13]. Several pretenders exist on what a smart city *could* be (Songdo in Korea, Masdar in Abu Dhabi,…) but they are not cities to live in, they are demonstrators, propelled by big companies (e.g. Cisco in Songdo) who apply a particular technology to the conception of a city. In the literature, the smart city is recently defined as an ecosystem, that is to say a system where the whole is more than the sum of the parts and has autopoeitic properties [14].

For the system architect this approach implies:

- Defining a perimeter that comprehends all the components that have a critical impact on city life: the city needs to be fed, is to import products that may have been manufactured on a basis that does not necessarily fit with sustainable development requirements (pollution, children work or underpaid workers, carbon emissions…). These costs and environmental impact must be charged to the city balance.
- Considering the system as a living system where the behavior of inhabitants determines the sustainability of the ecosystemic properties of the city. The underlying assumptions are material systems in addition to immaterial ones—as history, culture, anthropology and social capital—play their role. A recent trend in the literature on development economics, which is contrary to the fad of mainstream economics that consider all territories alike, put the emphasis on the "smart territory" as an unstructured cluster of tradition, culture, and informal institutions able to shape an innovative milieu [3].

Assuming the city is an ecosystem, according to the laws of general system theory [2] it may be conceived as shown in Fig. 1:

(a) **Finality**: It has a finality made of strategic vision borne by stakeholders (public and economic actors), people living in the city and sustaining this finality

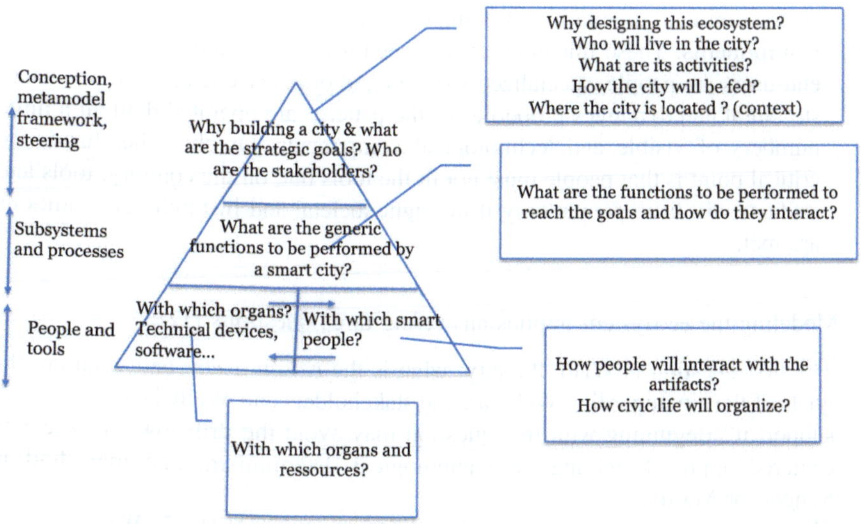

Fig. 1 Architecting the ecosystem

through theirs activities, and preserves its identity by interactions with its environment.

(b) **Functional tree structure**: This system may be broken down in tree structures of subsystems: the functions. These functions belong to hard and soft domains (Fig. 4). Hard domains include energy, water, waste, transport, environment, buildings, and healthcare infrastructures. Soft domains include education, welfare, social capital, public administration, work, civic activity and economy. What makes the city intelligent is the richness (quantity and speed) of connections between branches. We speak of a tree structure in the sense of Herbert Simon's architecture of complex systems [21] where the designer connects the subsystems to make the system emerge according to the aim it pursues. In his seminal paper "a city is not a tree" (1965) Christopher Alexander, an architect initially trained as mathematician and Professor at Berkeley, criticized the conception of the urban planning movement in America, considering it as a "fight against complexity", with no connections between branches. Modern cities conceived for cars, compared to ancient cities, offer a very poor web of connections.

(c) **Patterns**: Alexander formalized his idea of the city conceived as a rich overlapping of building blocks in his 1977 book *A pattern language*. This insight of considering the whole as a combination of modular and reusable building patterns (referring to structures, objects and events), lingered on the margins of cities architecture but has had an enormous influence in the development of object oriented architecture in software design. Architecture patterns can incorporate practices that have proven successful in the past. This importance of patterns is today recognized in system design with Pattern based Systems Engineering (PBSE). Patterns provide a common language independent from the underlying technology that may be used at different levels of abstraction and granularity (Broodney 2014).

(d) **Components**: These functions are operated using tools and artifacts of which end-users are people, specialized workers and ordinary citizens. On one hand, structural and dynamics properties of the patterns are operated thought a finite numbers of visible and technological components. On the other hand, the critical point is that people must not fit the tools but, on the contrary, tools and artifacts will fit to people only if the right societal and institutional conditions are met.

Modeling the ecosystem implies answering three questions [12]:

- The first question is WHY the city: what is the *raison d'être* and what are the goals of the city regarding WHO are the stakeholders and WHICH activities will support it? Beginning with this question may avoid the drift towards a techno centered approach relying on technological determinism, one may find in Songdo or Masdar.
- The question "why" is then deployed in questions WHAT: What are the functions the smart city must perform to reach these goals? These functions are

designed in processes grouped in subsystems aligned with the goal of the main system.
- The third set of questions concern HOW these functions will be processed by technical organs operated by the people who are the city executives and employees, and the city dwellers as end users.

2 The Global Framework: Urban Lifecycle Management©

We assume the rules of complex system modeling and system architecture apply to the city as well as they apply to products through PLM (Product Lifecycle Management) in that case according to a framework we call *Urban Lifecycle Management*© (ULM). The difference is a city never dies and must permanently renew its economic and social fabric as well as its infrastructure. An unsmart city will continuously expand according to the laws identified by West et al. [24] that reveal increasing returns in infrastructure investment that allow the city to sprawl indefinitely. The complexity will grow out of control, resulting in a city being the sum of heterogeneous boroughs with strong social and economic heterogeneity and spatial dystrophy.

We define ULM first and foremost as a tool to design an ecosystem which will be coherent with the political, social and economic goal people assign to the city according to the principle of sustainable development: stability, waste recycling, low energy consumption, and controlled scalability, but in a way that allows to foresee its evolution and to monitor the transition in different ages of the city. ULM has to counterweight the appeal of technological determinism: in the past, technologies have always dwarfed their intended design and produced a lot of unintended results (Townsend 2013). ULM has to monitor the life of the smart city alongside its evolution, as represented in Fig. 2

- **Cycle 1: Conception**. A city can't be thought out of its historical and cultural context represented by the territory of which the city is the expression. The smart city embarks a strategic vision based on a strategic analysis of the context and material and immaterial assets of the territory (GREMI 1986). The smartness of a city profoundly relies on what has been coined as "social intelligence" by prof. Stevan Dedijer in the years 1970s as the capability to build consensus where each social actor relies on others to create new knowledge. Intelligence doesn't operate in a vacuum but is socially and culturally rooted [5].
- To be livable, the city may not be a prototype: the system architect must focus on the task of integration that needs, to be reliable, to proceed from off-the-shelf components that already have an industrial life and may be considered stable and reliable, in the same way the classical architect does not invent the brick in

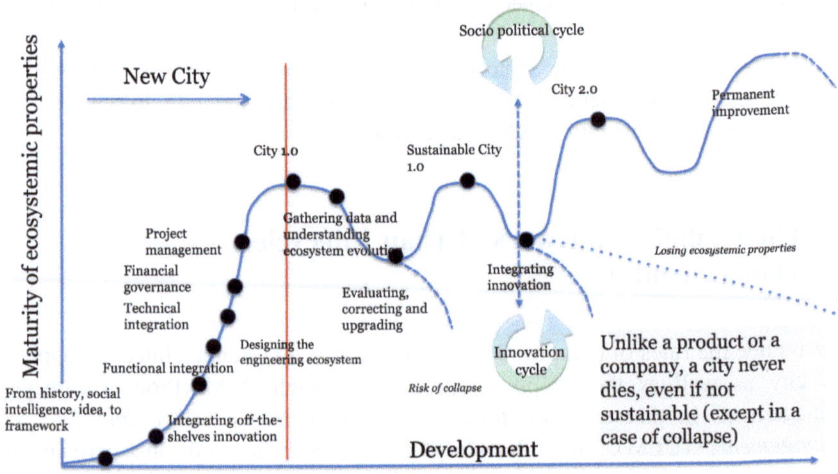

Fig. 2 Urban lifecycle management©

the same time as he designs the house. This will imply coordination between innovation cycles as we will see further.

- **Cycle 2: Datafication**. The process carried out on the principles represented in Fig. 1 leads to a first release of the city 1.0 in case of a new city. Just as well in a new or old city, we need to understand how the city lives and the unavoidable discrepancies between intended design and real result, an observatory must be implemented that will collect data produced by the city. These data are of two kinds: (a) historical data that help understand the path dependency of the city, and (b) big data produced by the city daily life to understand how it lives. Corrections are made according to classical principles of quality process management.
- Alongside the lifecycle, exogenous innovation will occur that will need to be endogenized by the model. For example, Songdo in his initial design relied on RFID devices to track city dwellers. Today, smart phones have become the Swiss knife of the city dwellers, rendering the use of RFID devices obsolete. Innovation is ubiquitous in all subsystems of the city. Innovation in smart cars interacts with the architecture of transportation (hard subsystem) as well as in human behavior (soft subsystem). Innovation in the building blocks has very different lifecycles. Coordination will be needed through common frameworks such as projects management office extended to the global smart city's complexity.
- **Cycle 3: Innovation**. Innovation challenges the equilibrium of the smart city in two ways. First, disequilibrium may come from an innovation within a subsystem, which interaction with other subsystems must be tested to avoid unintended consequences, that precisely requires mastering the rules of system integration. Considering a city is an open system, theses rules won't ever be

finite and will need to be upgraded permanently. Second, not all innovations are compulsorily good for the city: Civic and political life have to evaluate the consequences of an innovation and to frame it so that it fits with the common good and the sustainability of the city.

- **Cycle 4: Continuous improvement**. All along its lifecycle, the city may lose its smartness with two undesirable consequences: the city may continue to sprawl on a non-sustainable basis leading to today clog cities. In case of a disruption in its core activity, the city may collapsed as it happened in the past when things had become too complex to be monitored, as studied for past civilizations by archeologist Tainter [22]. Reducing the size of the city is then the only solution to reduce the complexity. A similar thing appears today in Detroit, a city that has lost its goals and population, leading to the decision of reducing the size of the city as the only means of avoiding bankruptcy of an unmanageable and unproductive city. A similar pattern exists with the Russian monocities [11].

3 A Research Program: The Rationale for Urban Lifecycle Management (ULM)

ULM is based on the assumption that common rules of modeling may be defined, consisting in three main principles.

3.1 Strategic Analysis

As represented in Fig. 1 the first task is to define the issues with the stakeholders. The functions needed to reach these issues are then defined (Fig. 3), and deployed in organs and specific competencies and resources.

3.2 Inventorying the Building Blocks

In spite we may define general rule of modeling, the smartness of a city will always be specific to the context, e.g. geographical and climate constraints (a city exposed to tropical floods or earthquake will embark functions that a city in a temperate country won't need), economic activity (specialization, search for synergies, position on the commercial routes and worldwide supply chains). The selection of these functions is essential to build a resilient city, e.g. with the climate change new phenomenons occur such as flood, marine submersion, extreme frost, heat waves

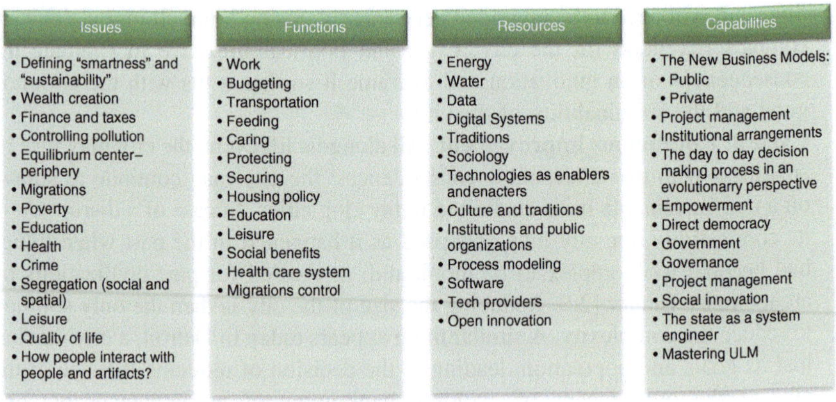

Fig. 3 The building blocks

the city was not prepared for. Nevertheless, common functions will exist in every city and their organization may proceed from off-the-shelf patterns.

3.3 *Integrating the Ecosystem*

In complex systems dynamics, the behavior of a system as a whole is an *emergence*, that is to say the property of the system can't be attributed to one function in particular but is the result of interactions between these functions. The "good life" is the basic question of political philosophy since Aristotle. It is an ethical issue that will result from political and strategic debates among the stakeholders. Jacobs [10] had criticized the utilitarian approach that prevailed in America in the city planning movement. The ancestor of the urban planning movement, Ebenezer Howard, thought of the smart city as an ideal city conceived from scratch as a mix of country and city. His insight was to conceive the city as an interaction between a city with jobs and opportunity but with pollution, and the countryside with fresh air and cheap land but with fewer opportunities, each one acting as magnets attracting and repelling people. He invented a third magnet, the Garden city, which combined the most attractive elements of both city and countryside [9]. Garden city was the Songdo of its day (Townsend 2013) that galvanized architects, engineers and social planners in search of a rational and comprehensive approach of building city. Howard's approach was excoriated by Jane Jacobs in his *Death and Life of Great American Cities* (1961) for not giving room to real life: "He conceived of good planning as a series of static acts; in each case the plan must anticipate all the needed… He was uninterested in the aspects of the city that could not be abstracted to serve his utopia". In fact, the city garden dream, not relying on a global systemic architecture, has degenerated in the banal reality of suburban sprawl.

The same risk exists today with digital technologies, which could revive the ideal city dream, under the impulse of the big players such as Cisco, IBM, Siemens, GE who have interest in a top-down and deterministic approach that reduce smart cities to the adoption of their "intelligent" technology. To avoid this bias system architecture must focus on four points:

(a) **Soft and hard subsystems**: Today's prototypes of would be smart cities are techno driven but mainly forget the inhabitants. City dwellers have the main role to play since it is their behavior and their use (and more and more the production) of information and technology that make the day to day decisions that render the ecosystem smart or no. Figure 4 represent both parts of the ecosystem the *soft* one, or human subsystem, and the *hard* one, the group of technical subsystems. Integration of these subsystems obeys different laws: human subsystems are dissipative ones, difficult to model, not obeying physical laws, with important entropy. Reducing their uncertainty relies on the sociology of uses, social consensus based on accepted formal and informal institutions, and a close association of inhabitants to the design of the system, which is a common feature of complex system design. Physical subsystems are conservative ones that can be modeled through the laws of physics with a possibility to reduce entropy, but keeping in mind that the decider in last resort is the city dweller who will use it.

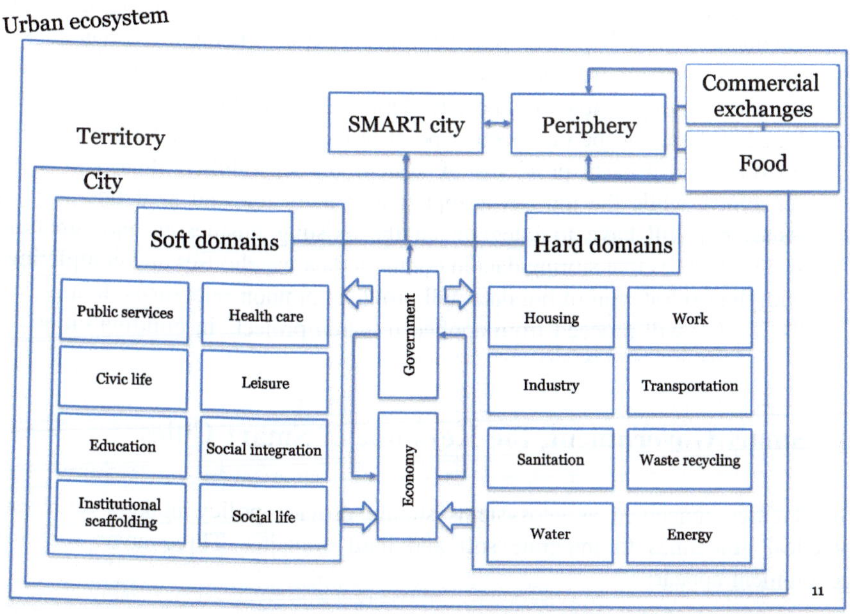

Fig. 4 The smart city as an emergence

(b) **Outside/inside**: The urban ecosystem is not reducible to the city itself, with perhaps the exceptions of city-states like Singapore. A city must be fed and have exchanges with a close periphery which produces goods (services, agriculture, food…) in interaction with the center. The design of a system relies on the definition of its border. According to the laws of requisite variety (Ashby law) the inner complexity of a system must be appropriate to the complexity of its environment. So, the urban ecosystem will have to define three perimeters: the *first* is the city itself where the synergies and interactions are the stronger and have the most "eco" properties. The *second* is the periphery: one may refer here to the model defined by Thünen at the beginning of the 19th century representing the city with a succession of concentric rings going from the highest increasing return activities at the center city to decreasing return activities at the periphery [20]. The *third* is the external environment with witch the city exchanges, that is, in a age of a globalized world, the rest of the world: the larger this perimeter, the more the system exchanges. This represents logistic costs that may have a negative impact on pollution and carbon emission that may be reincorporated in the balance of the city to measure its smartness, and the more it is subject to external factors of instability and the lesser the ecosystem is coherent and stable as a Thünen zone.[1]

(c) **Combining top down and bottom-up integration**: Each industry has today its model for the integration of its activities. Smart grids, water suppliers, transport operators, IT providers … have model for systemic integration of their subsystem and to evaluate its impact on the global functioning of the city. On the other hand, we know that the urban ecosystem being more than the sum of the subsystems we need another approach that starts from the top, that is from the strategic goals of the city deployed in functions as represented in Fig. 1. Where will be the meeting point of these two approaches? Proceeding bottom-up will raise problems of system interoperability, data syntax and semantics, while the top-down approach is more relevant to define strategic issues but will have to integrate all the existing businesses and functions. A possibility is that storing data in common data warehouses and completing it with the exploitation of big data will provide common references. In any case, the answer will proceed from applied research projects in building cities.

4 Smart Government, the Keystone of Smart Cities

Smart cities conceived as ecosystems should provide policy makers with some practical guidelines to integrate soft and hard domains. Three areas for smart government appear:

[1]We may give as an example the city of Quimper at the heart of the granitic massif of Brittany (France) who choses to import its granite from China.

Economic development: In the past, smart cities have been built without central planning (except in the case of Roman cities which reflected the imperial objective of the Roman Empire) but with a clear, although not explicitly formulated, founding purpose: defense, commerce, religion, power, geography… The pattern of the city emerged out of the interactions of key stakeholders: The lord, the barons, the merchants, the shopkeepers, the craftsmen, the bankers and the people. The design of ancient cities made them intelligent since they were ecosystem that sustained and reinvented themselves along time… till the point their capacity to self-reinvent came to an end when the core of their strategic activity reached a tipping point (e.g. Italian cities after the Renaissance, Russian mono-cities from the USSR era, Detroit today). The design of these cities obeyed to the real interactions underlying economic life (roads, markets, fairs, harbors, work, industry…) and civic activities (agora, city hall, structure of power). The task of government is to search for the activities that produce the highest increasing returns, no thanks to high technology but to synergies between activities (Reinert 2012), that will constitute the center of the Thünen zones. The Russian mono-cities built on a unique industry (coal, oil, cars, aerospace…) linger as long as this industry has a leading role but have very poor capabilities to reinvent itself due to the lack of synergies between different economic activities.

A vibrant political life: With cities emerged political philosophy. The most perspicacious analyst of what makes a city great was undoubtedly Machiavelli who put emphasis on the necessity of the common good: "*it is the common good and not private gain that makes cities great*" he wrote in his Discourse on Livy. Machiavelli conceived the common good in the Thomas Aquinas' tradition as a whole superior to the sum of its parts. Its systemic equilibrium is permanently challenged by the corruptive forces of *fortuna* that must be offset by the *virtù* of the Prince and the dynamism of the *vivere politico* [19]. Emphasis has been put on the topicality of Machiavelli to understand the systemic character of public management [18]. The vitality of the system is sustained with permanent interactions within thanks to a vibrant political life that provide a space for controversies. Machiavelli praised the Roman republic for his institution of the tribunate that managed the confrontation between the many of the citizens and the few of the ruling class that allowed the Republic to upgrade his institutions according the principles of the common weal advocated by Cicero. In contemporary complex societies, Elinor and Vincent Oström have developed the concept of polycentric governance that is organizing governance on one hand on a vertical axis from upper to lower levels of complexities, and on the other hand on an horizontal axis which consists of overlappings between organizations [16]. Elinor and Vincent Ostrom have criticized the excess of rationality that defines strict boundaries within missions and attributions of public organizations, since the reality doesn't know these boundaries and the adaptive character of public systems may be found in their overlaps.

Supporting open innovation: The experience of cities opening their database to the public to trigger the development of apps has proved the payoff of bottom-up approaches: in Washington DC, a contest "apps for democracy" challenged the local developers to create software exploiting public resources. For a cost of 50 000

US$ the pay-off was blazingly fast with forty seven apps developed in thirty days, representing an estimated 2 million worth of services, about 4000 % return on the city investment (Townsend 2013). But one should not conclude that bottom-up approaches are the killing solution: theses apps are V 1.0 developed by techies on the basis of a fascination for technologies while the city needs V 7.0 tested and reliable and based on the real needs and problem solving of citizens as end-users not familiar with technology. We rediscover here one of the law of innovation emphasized by Von Hippel [23]: the key role of lead users in the innovation process which is furthermore not a specific aspect of innovation in the digital era but a permanent, although forgotten, feature of the innovation process in the industrial era as reminds us François Caron, a leading academic in history of innovation [4].

In the same manner national innovation systems exist [6] and provide a framework that gives incentives to cooperation between industry, research and investors to steer their activities toward risk taking innovations, extended public administration could structure an urban innovation system that would structure the innovation process in a way that would guarantee that innovation, research and development of so-called smart apps are focused on the real needs of the city dwellers.

This approach requires a combination between soft and hard domains that can be achieved through complex systems of systems (SoS) architecture [7], a new discipline, methodology and competency we coin as urban lifecycle management©. The newborn concept of extended administration finds here its application in its intention to encompass and to design the global value chain of public administration and its interaction with—and between—all the stakeholders. This implies a sea change in the competencies and business model of public administration. This new field would be carried out through research in action projects building cities as ecosystem tending toward resilience where humans are first to decide for the ends.

References

1. Alexander, C.: A pattern Language, Town, Buildings, Constructions, with Sarah Ishikawa et Murray Silverstein. Oxford University Press, Oxford (1977)
2. Ashby, W.R.: Principles of the self-organizing system. In: von Foerster H., Zopf G. (eds.) Principles of Self-Organization, pp. 193–229. Pergamon, Oxford, Cambridge (1962)
3. Aydalot, P. (ed.): Milieux Innovateurs en Europe. GREMI, Paris (1986)
4. Caron, F.: La dynamique de l'innovation. Albin Michel, Paris (2012)
5. Dedijer, S.: Au-delà de l'informatique, l'intelligence sociale. Stock, Paris (1984)
6. Freeman, C.: The national system of innovation in historical perspective. Camb. J. Econ. **19** (1), (1995a)
7. Godfrey, P.: Architecting complex systems in new domains and problems: making sense of complexity and managing the unintended consequences. In: Complex System and Design Management, Proceedings (2012)
8. Hardin, G.: The tragedy of the commons. Sci. New Ser. **162**(3859), 1243–1248 (1968)
9. Howard, E.: Garden Cities of Tomorrow, 2nd edn. S. Sonnenschein & Co, London (1902)
10. Jacobs, J.: Cities and the Wealth of Nations. Random House, New-York (1985)

11. Kirsanova, N.Y., Lenkovets, O.M.: Solving monocities problem as a basis to improve the quality of life in Russia. Life Sci. J. **11**(6), (2014)
12. Krob, D.: Eléments d'architecture des systèmes complexes. In: Appriou, A. (ed.), Gestion de la complexité et de l'information dans les grands systèmes critiques, pp. 179–207. CNRS Editions (2009)
13. Lizaroiu, G.C., Roscia, M.: Definition methodology for the smart cities model. Energy **47**, 326–332 (2012)
14. Neirotti, P., De Marco, A., Corinna Cagliano, A., Mangano, G., Scorrano, F.: Current trends in smart city initiatives: some stylised facts. Cities **38**, 25–36 (2014)
15. Ostrom, E.: Governing the Commons; the Evolution of Institutions for Collective Action. Cambridge University Press, NY (1991)
16. Ostrom, E.: Beyond markets and states: polycentric governance of complex economic systems. Am. Econ. Rev. 1–33 (2010)
17. Reinert, S.A., Serra, A. (ed.): A Short Treatise on the Wealth and Poverty of Nations (1613). Anthem Press, London (2011)
18. Rochet, C.: Le bien commun comme main invisible: le legs de Machiavel à la gestion publique. Rev. Int. des Sci. Admin. **74**(3), (2008)
19. Rochet, C.: Qu'est-ce qu'une bonne décision publique? Editions universitaires européennes (2011)
20. Schwartz, H.: States Versus Markets: The Emergence of a Global Economy, 3rd edn. Palgrave, London (2010)
21. Simon, H.A.: The Sciences of the Artificial, 3rd edn. MIT Press, New York (1969)
22. Tainter, J.: The Collapse of Complex Societies. Cambridge University Press, Cambridge (1990)
23. Von Hippel, E.: Lead users: a source of novel product concepts. Manage. Sci. **32**(7), 791–806 (1986)
24. West, G., Bettencourt, L.M. A., Lobo, J., Helbing, D., Kühnert, C.: Growth, Innovation, Scaling, and the Pace of Life in Cities. Indiana University, Bloomington (2007)

Designing Systems with Adaptability in Mind

Haifeng Zhu

Abstract Designing a complex cyber-physical or manufactured system requires a significant amount of effort. A good design needs to be adaptable to requirement changes, however should also avoid unbounded margins that can be costly. Achieving this fine balance is difficult. This paper presents a design process that takes adaptability into consideration. By exploring the missions a system can support within a specified limit of additional engineering costs, we are able to characterize this system's adaptability. Such a characterization inherits the original meaning of adaptability in ecosystems that describes a system's ability of maintaining the original goals even when facing ongoing changes, and allows it be computable in industry. A new design process for a product family is then established to identify designs that support the most missions while controlling costs. An HVAC example is used to illustrate such a design process that helps maximize mission performance and reduce costs.

1 Introduction

Designing a complex system is costly and the designers often have to decide how much margin to reserve in all aspects, including the architecture level, to cater to the potential requirement changes from customers. It is sometimes difficult to make these decisions without quantitative measures.

This research was partially developed with funding from The Defense Advanced Research Projects Agency (DARPA)/The Air Force Research Laboratory (AFRL). The views, opinions, and/or findings contained in this article/presentation are those of the author(s)/presenter(s) and should not be interpreted as representing the official views or policies of the Department of Defense or the U.S. Government. Distribution Statement "A" (Approved for Public Release, Distribution Unlimited).

The original version of this chapter was revised: The spelling of the author's name was corrected. The erratum to this chapter is available at DOI 10.1007/978-3-319-26109-6_40

H. Zhu (✉)
United Technologies Research Center, East Hartford, USA
e-mail: ZhuHF@utrc.utc.com

G. Auvray et al. (eds.), *Complex Systems Design & Management*,
DOI 10.1007/978-3-319-26109-6_20

A lot of research exists in the area of flexible and reconfigurable manufacturing systems where more than 50 different flexibilities and relevant measures were studied [1–4]. Different kinds of modeling and measures [5–8] utilized different concrete techniques from decision theory, petri net, information theory, multi-dimensional approaches, etc. Flexibility may be valuable however requires upfront investment and justification that decision makers generally find difficult. Adaptable designs, however, focus on designing a new system from existing ones, thus allowing new capabilities to be supported later.

System architecture is important for both software and hardware material domains. A poorly chosen architecture may result in significant difficulty in supporting new requirements and missions. It is important for a designer to pick an architecture with good adaptability at the beginning; however, the traditional decision process (Fig. 1) cannot take this into account quantitatively.

To allow adaptability be evaluated in the early stages of system design, engineered system adaptability must be defined. Adaptability traditionally comes from ecosystems [9], which indicates the ability of a system or process to change something or oneself to fit to occurring changes, such as an unexpected disturbance in the environment [10] formulates it as: Given a system *S* that suffers a change due to a stimulus event *E*, S is an adaptive system if and only if the long-term probability that the system *S* change its behavior (S $\rightarrow$ *S′*) is same with and without *E*, which is:

$$\lim_{t\to\infty} P_t(S \to S'|E) = \lim_{t\to\infty} P_t(S \to S') \quad (1)$$

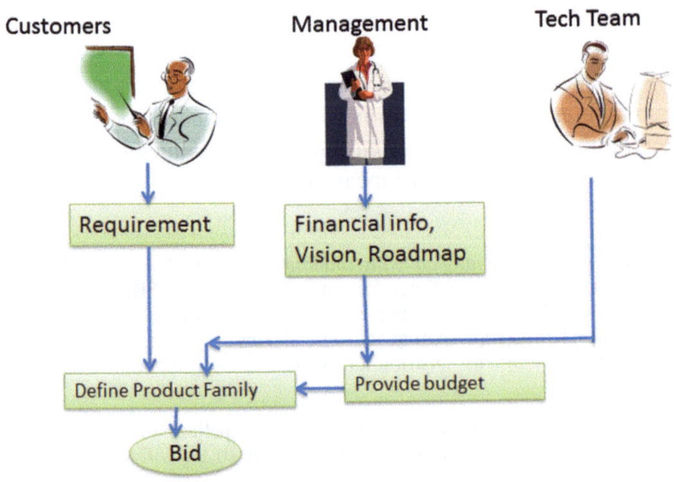

Fig. 1 Conventional decision process

In previous approaches in the system design area, different researches have different definitions that do not always inherit the meaning of adaptability in ecosystems. Some of them miss important elements of the concept, and some others focus on concrete modeling techniques that introduce applicability restrictions from these techniques. For example [3, 11] do not capture how to model the ongoing changes. Without defining missions, arguing how adaptable a product is by quantifying the engineering cost of producing another set of products from the current product state is not of much use, because a product can, with small cost, switch to a large number of other products that are not able to serve your original goals or even useless in reality. This is inconsistent with the original meaning of adaptability in ecosystems. Our work overcomes the above problems by modeling potential changes with a mission space and evaluating the adaptive capabilities of product architectures in a product family by their supports of the missions. These architectures are typically generated by design space exploration that exhaustively produces all possible valid architectures in the family. Each of these architectures can be evaluated with its adaptability metric, and a new design process taking this into account can be created. This way architectures with good adaptabilities can be identified, enabling the product to be highly adaptable to customer requirements or market changes. This is significantly useful in a competitive economic environment.

The remaining parts of this paper are as follows. Section 2 describes a preliminary exploration of using mission space to characterize system changes and possible formulation of an adaptability metric. Section 3 describes a synthetic example for HVAC systems. Finally Sect. 4 concludes our study.

2 Modeling Issue Discussions

The idea of a new design process is to include the evaluation of adaptability metric for each product architecture and combine this information with other factors in the decision process. To achieve that, let us first explore the modeling issues with adaptability.

Different systems may have different missions to accomplish. It is possible that we can use a set of missions to describe what a system is required to accomplish now and may be required to accomplish in the future. To explore what these possible missions are one may obtain them by enumerating combinations of all possible mission segments or tasks. However, we want to point out this is incorrect, because this leads to meaningless combinations. One possible way to perform the modeling is to use a state machine.

Figure 2 is an example of a set of states of a plane during an ordinary flight, which can be called a *mission state machine*. Formally, a mission state machine $\mathcal{M} = <m, e>$ where m is a set of *tasks* or *mission segments* to be performed by the system (i.e. states), and e is a set of directional transitions denoting conditions and requirements from states to states. From a mission state machine, one can

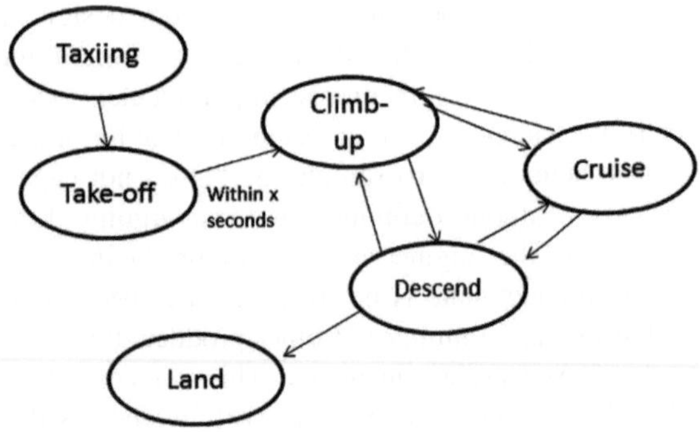

Fig. 2 Mission state machine

describe a *mission M* as a sequence of tasks (subset of *m*) connected via edges (subset of *e*) in the machine, which is a trajectory of the graph. Only a trajectory instead of arbitrary combination of tasks can be considered as a mission. For example: $M_1 = \{m_1, m_2, m_3, m_4, m_5, m_6\}$ where m_1 is taxiing, m_2 is take-off, m_3 is climbing at 1000 ft/min, m_4 is cruise, and m_5 is descend, m_6 is landing. Then, we can define a *mission evaluation space*, $\Omega = \{\mathcal{M}, <M_j, \xi_j>\}, j = 1, \ldots, |\Omega|$ where $|\Omega|$ is the cardinality of $<M_j, \xi_j>$ which is a set of all meaningful missions M_j associated with their properties ξ_j that both built on mission state machine $\mathcal{M}$.

Based on the above, the adaptability metric of an architecture can be defined based on how difficult it is to support all the required missions and maximum number of optional missions within certain switching cost, i.e. the design cost of another architecture based on the current one. Techniques such as those developed in [12, 13] can be used to calculate this cost. In this paper, we only consider development cost but other costs such as manufacturing and operational cost can be included under the same framework. An architecture's support to a mission can be formulated as an indication function in the form of deterministic or fuzzy logic membership function, for example $S(x)$ where x is a mission. $S(x) \in [0, 1]$ where 0 means not support, 1 means fully support, and any number between 0 and 1 means partial support based on its extent of support. Then the adaptability metric can be described using a utility function having values on the interval [0, 1] with different categories within each category a simple linear function can be used. For example: an architecture can be considered as Perfectly Adaptable if it fully supports all the missions with 0 additional cost. We may assign the adaptability metric to be 1 for this case. Otherwise, if it supports all the missions within reasonable amount of switching cost for additional engineering (i.e. within a user-specified threshold), it maybe termed as Mostly Adaptable, and obtained an adaptability value between 0.5 and 1. The less switching cost to support all the missions, the bigger adaptability value is assigned. Otherwise, a Partially Adaptable architecture supports, within user-specified cost threshold, all the required and only some of the optional missions (i.e. its summation of the S() function over the optional missions in mission evaluation space is less than the total number of optional missions). The adaptability metric can take the remaining of the interval except 0, with higher values when more optional missions are supported. Finally, an architecture supports only the required missions is Non-adaptable with an adaptability value 0.

3 New Design Process

With adaptability in mind, we can now define a new design process in the case of a product family (Fig. 3). Upon receiving customer's requirement, the tech team analyzes them and defines a mission space with all possible missions now and in the future. The tech team also enumerates all possible architectures in this product family. On receiving acceptable cost threshold from the management, the adaptability of each architecture can be calculated. The management and the tech team can then sit together and select an adaptable architecture as the basis for their product family.

As an example, we illustrate a cyber-physical system product family with two HVAC systems. System A is a single-zone system with a compressor, a condenser and an evaporator connected through a loop of pipes. A simple IC controls it and

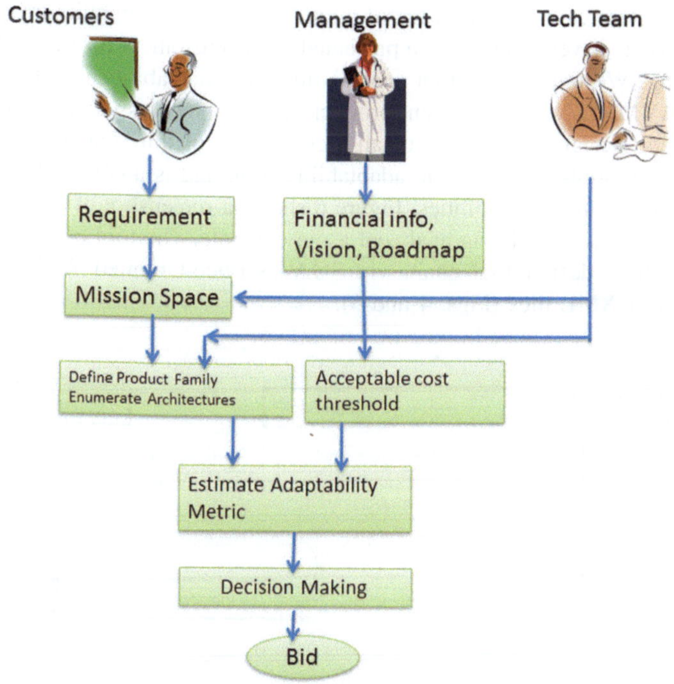

Fig. 3 Decision process taking adaptability into account

setpoints are fixed and burned into the IC. System B is a two-zone HVAC system with two dampers each of which connects to pipes for individual zones. In addition, it has a centralized zone controller that connects with the user control panels in each room via Ethernet link, which enables user control.

Let us assume we have three missions/requirements in the mission space: M_1 supports each room with fixed temperature at 20°, which is required. M_2 supports independent switch on/off for each room and fixed temperature at 20° when it's on, which is optional. M_3 allows the temperature of one of the rooms to be adjustable based on personal comforts, which is optional. The optional missions reflect customer's potential needs that are currently uncertain. Apparently System B can support all the missions, while system A can only support M_1 and M_2. Therefore, the adaptability metric for B is 1. Let's assume the switching cost from A to B is $C_{sw}(A, B) = 300$ K USD and our acceptable switching cost threshold is 200 K USD. A's adaptability falls into (0, 0.5) and is assigned with the medium point 0.25, as it supports only half of the optional missions within tolerable switching cost threshold. If the development cost of B is acceptable by the customer, the manufacturer should persuade the customer to select B during bidding. This is a simple illustrative example with only two architectures. In reality the design processes for complex systems, such as design space exploration, typically generate a lot of architectures/designs and the mission state machine can be very complicated (for example military cases). Thus, a list of all architectures' adaptability metrics along with their costs is very useful to be presented to the customers for decision making and bidding. We implemented an analysis tool for adaptability in a tool chain in DARPA AVM [12] where different product families (such as jet engines, etc.) can be analyzed. Design information of all architectures of different product families from upstream tools flow into the adaptability tool, and is analyzed for their supported missions and adaptabilities. Figure 6 shows a screenshot of this tool, where jet engine designs were being analyzed as an example. The tool can communicate with the whole design tool chain via SOA (Service-Oriented Architecture), or simple text or XML files (Figs. 4 and 5).

Fig. 4 HVAC system A—single zone

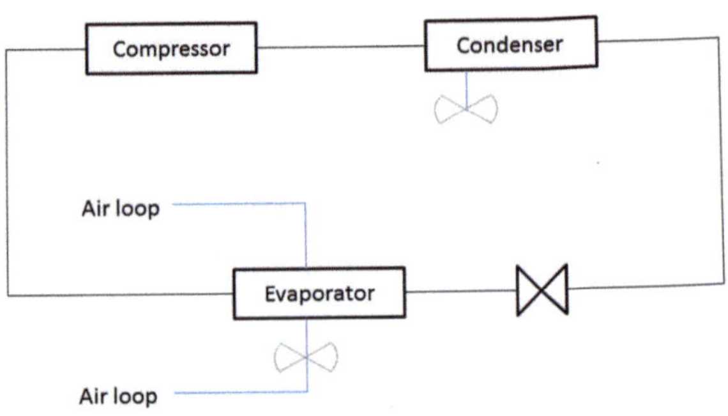

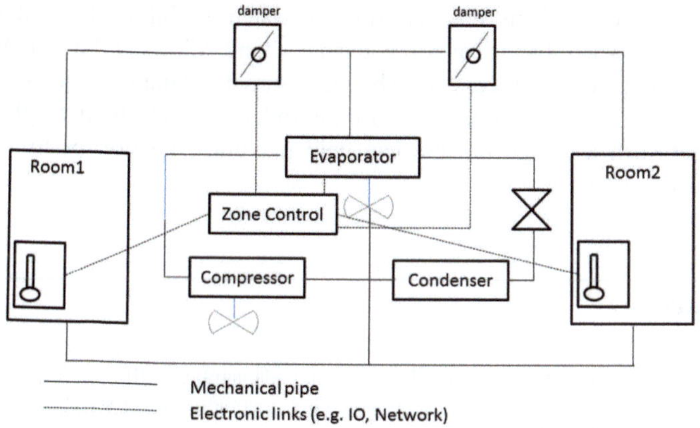

Fig. 5 HVAC system B—two zone

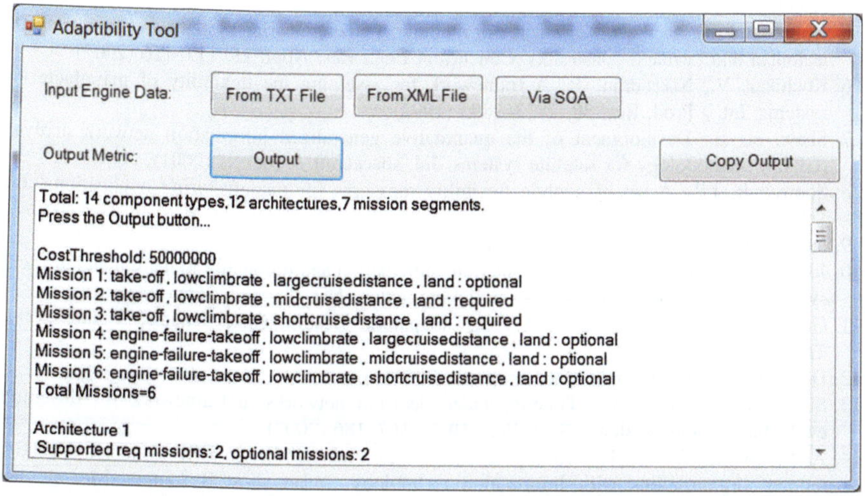

Fig. 6 Adaptability estimation tool

4 Conclusion

Designing a complex cyber-physical or manufactured system requires a significant amount of effort. A good design needs to be adaptable to requirement changes, however should also avoid unbounded margins that can be costly. Achieving this fine balance is difficult. This paper presented a design process taking adaptability into consideration. By exploring the missions a system can support within a specified limit of additional engineering costs, we were able to characterize this system's adaptability. Such a characterization inherits the original meaning of adaptability in ecosystems that describes a system's ability of maintaining the original goals when facing ongoing changes, and allows it be computable in industry. A new design process was shown for product families to identify designs that support the most missions with cost controlled. A synthetic example was also shown that such a design process can help maximize mission performance and reduce costs.

References

1. Chan, H.K.: Comparative study on flexibility and adaptability (2010)
2. Nilchiani, R.: Measuring space systems flexibility: a comprehensive six-element framework. Ph.D. thesis, MIT (2005)
3. Ross, A., et. al.: Defining system changeability: reconciling flexibility, adaptability, scalability, and robustness for maintaining system lifecycle value. INCOSE (2007)
4. Sethi, A., Sethi, S.: Flexibility in manufacturing: a survey. Int. J. Flex. Manuf. Syst. **2**(4) (1990)
5. Holtta-Otto, K., de Weck, O.: Degree of modularity in engineering systems and products with technical and business constraints. Concurrent Eng.: Res. Appl. **15**, 113–126 (2007)
6. Kochikar, V., Narendran, T.: A framework for assessing the flexibility of manufacturing systems. Int. J Prod. Res. **30**, 2873–2895 (1992)
7. Shaw, et. al.: Development of the quantitative generalized information network analysis (GINA) methodology for satellite systems, Jnl. spacecraft & rockets (2001)
8. Shewchuk, J.P.: A set of generic flexibility measures for manufacturing applications. Int. J. Prod Res. (1999)
9. http://en.wikipedia.org/wiki/Adaptability
10. Jose Antonio Martin H., et. al.: Adaptation, anticipation and rationality in natural and artificial systems: computational paradigms mimicking nature. Nat. Comput. **8**(4) (2009)
11. Gu, P., Hashemian, M., Nee, A.Y.C.: Adaptabile Design, CIRP Annals. Manufacturing Technology (2004)
12. DARPA Adaptive Vehicle Make program (2011)
13. Silver, M., de Weck, O.: Time-expanded decision networks: a framework for designing evolvable complex systems. Syst. Eng. **10**(2), 167–186 (2007)
14. Andresen, K., Gronau, N.: An approach to increase adaptability in ERP systems: managing modern organizations with information technology. Information Resources Management Association International Conference, 2005
15. Sinha, K., de Week, O., A network-based structural complexity metric for engineered complex systems. IEEE International Systems Conference (SysCon), 2013

Part II
Posters

Analysis of the INCOSE Rules for Writing Good Requirement in Industry: A Tool Based Study

José M. Fuentes, Anabel Fraga, Gonzalo Génova, Jose Álvarez and Juan Llorens

Abstract The Requirements Engineering (RE) discipline has been promoted, implemented and deployed for more than 20 years through standardization agencies (ISO/IEC, IEEE) and national/international organizations (such as INCOSE). Ever since, despite an increasing maturity, RE remains a discipline unequally understood and implemented, even within the same organization. Problems found in current Systems Engineering projects with focus in RE could be mitigated using quality metrics in the process. Quality metrics aids in the process of writing good requirements by following a reference guide. INCOSE has promoted and published a guide for writing good requirements, with support of several industrial and academic partners. The more correct, complete and consistent a requirement is, the best performance it will have, and fewer errors will occur in system developments and operation. This paper presents a study where a set of the published INCOSE rules have been implemented in a tool for assessing requirements quality.

J.M. Fuentes (✉)
The REUSE Company, C/Margarita Salas, Legatec, Madrid, Spain
e-mail: jose.fuentes@reusecompany.com

A. Fraga · G. Génova · J. Álvarez · J. Llorens
Carlos III University of Madrid, Av. Universidad, Leganes, Madrid, Spain
e-mail: afraga@inf.uc3m.es

G. Génova
e-mail: ggenova@inf.uc3m.es

J. Álvarez
e-mail: jalvarez@inf.uc3m.es

J. Llorens
e-mail: llorens@inf.uc3m.es

G. Auvray et al. (eds.), *Complex Systems Design & Management*,
DOI 10.1007/978-3-319-26109-6_21

Implementing Model Semantics and a (MB)SE Ontology in Civil Engineering and Construction Sector

Henrik Balslev

Abstract In the period from 2010 to 2015, the Danish Building Construction Sector has implemented basic parts of Systems Engineering as the new 'common language' in the building construction sector. The project is anchored in the public and EU supported "cuneco project". www.cuneco.dk develops the common basis for digitalized cooperation in construction, operation and maintenance to increase efficiency and productivity through enhanced exchange of information. To allow maximum simplicity yet unlimited flexibility, systems and their constituents are first classified and then identified individually to be used consistently over the lifecycle of the component, suitable for IT support. The system-of-systems principle is a fundamental approach to achieve unambiguous identification based on the Reference Designation System principles as defined in ISO/IEC 81346 standard series, which originally is designed for modelling and labelling of any kind of industrial plant. Currently, the Danish result is used to update some parts of the 81346 standard series, and thereby introducing Systems Engineering to the building construction sector.

H. Balslev (✉)
Systems Engineering A/S, Østerbrogade 56A, DK-2100 Copenhagen Ø, Denmark
e-mail: hb@syseng.dk
URL: http://www.syseng.dk; http://www.81346.com

G. Auvray et al. (eds.), *Complex Systems Design & Management*,
DOI 10.1007/978-3-319-26109-6_22

E-vehicle Service Architecture for Logistic Systems

Sebastian Apel and Volkmar Schau

Abstract Until the year 2020, Germany has established a national development plan with the goal to push one million fully electric vehicles into use. Part of the plan is to establish a number of federally funded research projects, which investigate and tackle domain specific problems, e.g. the limited driving range of electric cars. Freight traffic is especially hampered by those range restrictions. The Smart City Logistik project (www.armor.uni-jena.de/www.smartcitylogistik.de) strives for a practical and short-term solution to this problem in the concrete context set by the city of Erfurt, Germany. The focus is on ICT-support for currently available, small and medium sized, fully electric vehicles that provide for the "last mile" in freight handling. This poster provides the first results of on going work to construct an architecture managing these requirements with a special focus on how to handle the wide range of interfaces.

S. Apel (✉) · V. Schau
Department of Computer Science, Friedrich Schiller University Jena, 07743 Jena, Germany
e-mail: sebastian.apel@uni-jena.de

V. Schau
e-mail: volkmar.schau@uni-jena.de

G. Auvray et al. (eds.), *Complex Systems Design & Management*,
DOI 10.1007/978-3-319-26109-6_23

EGNOS V3: Engineering the Future of GPS and Galileo Augmentation Over Europe

Jean-Alexandre Gicquel, David Arnaudy and Philippe Gouni

Abstract EGNOS provides today augmentation services based on GPS. It allows getting improved performances in a wide range of navigation applications, in particular for aeronautical approaches in the civil aviation domain. In parallel, GPS, Galileo and other constellations are evolving, and new services are identified to serve European users communities, answering to the emergence of new end-users applications needs, finally calling for the EGNOS V3 generation. In the area of navigation, EGNOS is designed to support Safety of Life applications, with stringent aeronautical performance requirements. In the same time, continuity of the EGNOS service to end users when evolving and security aspects of the solution shall be ensured, and furthermore, the solution is required to have improved operability and reduced lifecycle cost, that implies to pay specific attention in operations design. This paper provides an overview on how the Thales Model Based System Engineering (MBSE) methodology and tools are tailored and applied to support EGNOS V3 engineering objectives. A tooled-up environment is set-up to support concurrent engineering on a common design reference and to contribute to the consolidation and justification of the EGNOS V3 system architecture design and requirements. The resulting work organization and interactions between system engineering team and engineering domain specialists (safety, security, operations…) are presented. Finally, this paper is providing lessons learned and success stories of a model based approach to federate concurrent engineering activities, identifying the main outcomes and benefits.

J.-A. Gicquel · D. Arnaudy · P. Gouni (✉)
Observation Exploration and Navigation Business Line/Navigation Domain, Thales Alenia Space, Toulouse, France
e-mail: Philippe.gouni@thalesaleniaspace.com

G. Auvray et al. (eds.), *Complex Systems Design & Management*,
DOI 10.1007/978-3-319-26109-6_24

Integrating the ISO/IEC 15288 Systems Engineering Standard with the PMBoK Project Management Guide to Optimize the Management of Engineering Projects

Rui XUE, Claude Baron, Philippe Esteban and Li Zheng

Abstract As economic pressure continues to mount worldwide, cooperation between people, companies and even countries is becoming increasingly needed. At the same time, the scale of project is being revised upwards daily. In order to ensure the success of large scale projects, the manner in which cooperation is set up between different teams, such as systems engineers and project managers, is becoming an important issue. Cooperation between systems engineering and project management is now key in this respect. On the other hand, it is widely recognized that the use of standards can improve the success ratio. Thus, integration using standards or guides from systems engineering and project management can help companies improve their competitiveness. A host of standards or guides have already been published in both domains. The purpose of this paper is to choose those most frequently used standards or guides from the systems engineering and project management in order to compare and build a bridge between them and provide a view shared by systems engineers and project managers enabling them to carry out the project effectively.

R. XUE (✉) · C. Baron · P. Esteban · L. Zheng
LAAS, CNRS, 7 av. du col. Roche, F-31400 Toulouse, France
e-mail: rui.xue@laas.fr

C. Baron
e-mail: claude.baron@laas.fr

P. Esteban
e-mail: philippe.esteban@laas.fr

L. Zheng
e-mail: li.zheng@laas.fr

R. XUE · C. Baron · L. Zheng
INSA, LAAS, Université de Toulouse, F-31400 Toulouse, France

P. Esteban
UPS LAAS, Université de Toulouse, F-31400 Toulouse, France

G. Auvray et al. (eds.), *Complex Systems Design & Management*,
DOI 10.1007/978-3-319-26109-6_25

Taking Handicap into Account: Systemic Features

Patrick Farfal

Abstract The approach of handicap must resolutely be systemic. At least because the matter of handicap obviously and immediately addresses the question of social link, which is reciprocal by definition. Also because handicap as a fact is far from being marginal: one European out of ten is concerned by handicap; nearly 10 million of disabled persons (in a broad sense) can be counted in France. Only 15 % of disabled persons contract handicap at birth, so, any valid person may contract a handicap any day. Differences, also diversity, factors of complexity, demand a systemic approach. Lastly, handicap needs compensation (sensory or motor aid…, desk fitting out…): it is the environment which adapts itself to the disabled person!

In practice, and, generally speaking, in the society, individualism takes the lead over "living together". Stereotypes on disabled persons (deemed less performative, generating extra costs…) become widespread among people both in everyday life and at work. Answers provided by some elected members or administrators are not sufficient because they are fragmentary (for example limiter to training), while a set of consistent and complementary answers are needed.

The whole of those answers must include time factor; the point of view on disabled persons must be educated from childhood, from primary education. So, a systemic treatment of handicap implies coordinated actions in the following fields: children (welcome, education…), companies and employment (competences acknowledgement, recruitment…), administration (welcome and support, recognition of disabled worker status…), training (of disabled people, nursing staff, but also recruiting people and employers…), accessibility (to housing, buildings, transports, cultural and associative life, and of course cure and care), right to compensation (of sensory or motor handicap…). Even the component cure and care is of systemic nature: the person must be treated in her whole (therapeutic education, medicine acting at each step of the care path, care directed towards the transition to social and occupational rehabilitation, disabled person acting throughout her path…). Associations dedicated to handicap, who treat, educate,

P. Farfal (✉)
PatSys, 25 rue Jean Leclaire, 75017 Paris, France
e-mail: Pfarfal.patsys@sfr.fr

G. Auvray et al. (eds.), *Complex Systems Design & Management*,
DOI 10.1007/978-3-319-26109-6_26

train, insert, support, and those who, in their sports, cultural or artistic activities, include a handicap part, obviously play a major role in that approach.

Unexpected spin-offs of the compensation of handicap can be seen: the adaptability of some space (building, transport) to the needs and constraints of a person with a loss of autonomy is not a simple respect of law as regards accessibility, but is broadened to the quality of use of "life spaces" for everybody, taking into account the needs and constraints of the whole of people: the disabled person often appears to enlighten the needs of the whole (example: access platforms to busses). Considering system engineering vocabulary, that amounts to speaking of taking into account the needs and constraints of all the stakeholders, which is an essential condition of secure outcome of a project.

The adaptation of the environment to the disabled person, in the very scope of the February 11th 2005 French law, as well as the claim of her full citizenship (schooling, employment…), with its consequences onto the whole of people is not the least surprise arisen from thinking about handicap.

Considering the systemic features of the question of disability would make it possible for some elected or administration people not to immediately focus on solutions, often fragmentary, without any care of other relations between the actors of the field and their environment, but on the contrary tackle the question as a whole, and think about the benefits induced on "valid" people, major part of the population.

A Feedback Experience on DELTA SR: A Smart Tool to Compare Complex SCADE Models

Stéphane Fechter and Myriam Marchand

Abstract The signaling railway system company Ansaldo STS develops, with the formal language SCADE, a Carborne Controller for a SIL 4 CBTC (a management system for communicating urban trains). The Carborne Controller SCADE model is a critical software, embedded in the trains, of the CBTC system: 1026 SCADE operators to implement 1323 system requirements and 17 levels for the depth. To be compliant with the standard CENELEC EN 50128, Critical Code Reviews are mandated for the Carborne Controller SCADE model. Without support solution for Critical Code Reviews on complex SCADE models, we have developed a tool: Delta SR. Developed with TCL language, thanks to a heuristic based on textual, syntactic and semantic analyses, it computes a classification of differences between two SCADE models and exhibits the functional impacts of changes. The paper presents a feedback on DELTA SR and on its added value for the Critical Code Reviews on SCADE models.

S. Fechter (✉)
Safe River, 9 bis rue Delerue, 92120 Montrouge, France
e-mail: stephane.fechter@safe-river.com
URL: http://www.safe-river.com

M. Marchand
Ansaldo STS, 4 avenue du Canada, BP 243, 91944 LES ULIS Cedex, France
e-mail: myriam.marchand@ansaldo-sts.com
URL: http://www.ansaldo-sts.com

G. Auvray et al. (eds.), *Complex Systems Design & Management*,
DOI 10.1007/978-3-319-26109-6_27

A Systems Approach to Improve Performance in Supply Chain: Case Study in a Procurement Process in the Aeronautical Industry

Denis Olmos-Sanchez, Jean-Claude Bocquet and Marie-Agnès Forman

Abstract Supply Chains (SC) are becoming more complex by the interaction of various elements, and decisions must be taken at different levels to accomplish their objectives. Several approaches propose performance improvements but there is a lack of application of systemic approach to maximize the value creation. In this work, we apply a method called SCOS' (Systemic for Complex Organizational System) which focuses in reaching new objectives in terms of value creation (performances as economic, quality, time and environment) for each phase of the life cycle, and each stakeholder of the system (procurement process), then processes are developed to meet these finalities. A case study is used to model value creation in an SC as an improvement expected by stakeholders, and it is validated by industrial experts. Then recommendations are given to simulate and quantify these improvements through system dynamics.

Keywords Systemic approach · Modeling · Supply chain performance · Procurement · Improvement · Aeronautical industry

---

D. Olmos-Sanchez (✉) · J.-C. Bocquet
Laboratoire Génie Industriel, CentraleSupelec, Campus Châtenay-Malabry Grande voie des Vignes, 92295 Châtenay-Malabry, France
e-mail: denis.olmos@centralesupelec.fr

J.-C. Bocquet
e-mail: jean-claude.bocquet@centralesupelec.fr

M.-A. Forman
Dassault Aviation, 78 Quai Marcel Dassault, 92210 Saint-Cloud, France
e-mail: marie-agnes.forman@dassault-aviation.com

G. Auvray et al. (eds.), *Complex Systems Design & Management*,
DOI 10.1007/978-3-319-26109-6_28

CoDA—A Model-Based Platform to Deal with the Inherent Complexity of Automation Systems Development

Juan Navas, Patrick Herbert and Gilles Boussaroque

Abstract Automation Systems in AREVA are highly versatile, often reactive, systems that provide information treatment and control tasks to nuclear industry processes. These systems are inherently complex, as they involve many interconnected elements which behavior is not always well understood or predictable. Furthermore, they can also be considered as complex regarding their development process, as they demand a strong involvement of several stakeholders. The CoDA method and platform proposes a set of open and interoperable tools addressing Automation Systems' inputs Analysis, Design, Implementation and Verification and Validation activities. The integrated method and tools reduce time spent on impact analysis and provide proof of the proper consideration of requirements. This poster details the main propositions and results of the deployment of the CoDA platform in AREVA.

J. Navas (✉)
AREVA NP SAS, 1 Place Jean Millier, 92400 Courbevoie, France
e-mail: juan.navas@areva.com

P. Herbert
BP38 25 Avenue de Tourville, 50120 Equerdreville, France
e-mail: patrick.herbert@areva.com

G. Boussaroque
Euriware—Capgemini, 1 Place des Frères Montgolfier, 78280 Guyancourt, France
e-mail: gilles.boussaroque@euriware.fr

G. Auvray et al. (eds.), *Complex Systems Design & Management*,
DOI 10.1007/978-3-319-26109-6_29

Contingency Factors for Relationships in Complex Product Creation Environments

Donna Champion

Abstract Current approaches to systems design and management are at the limits of applicability in modern complex product design environments. The collaborative nature of design activity is increasingly difficult to manage, where multi-disciplinary teams must share knowledge and co-ordinate the integration of technologies across different platforms and architectures. This paper describes a qualitative study to explore the critical factors in building and sustaining relationships across cross-functional teams in complex product creation environments. The study was undertaken in the Automotive sector, where market pressures demand swift integration of new technologies across platforms. A number of contingency factors have been identified and three strategic priorities for managers are suggested.

D. Champion (✉)
School of Business and Economics, Loughborough University,
Loughborough LE11 3TB, UK
e-mail: d.champion@lboro.ac.uk

G. Auvray et al. (eds.), *Complex Systems Design & Management*,
DOI 10.1007/978-3-319-26109-6_30

Siting Nuclear Power Plants Incorporating Strategic Flexibility

Michel-Alexandre Cardin, Sizhe Zhang and William J. Nuttall

Abstract Nuclear power is an important energy source for generating electricity in consideration of CO_2 emissions and global warming. Siting nuclear power plants is a challenging issue nowadays due to the volatility of long-term electricity demand, as well as public acceptance of nuclear technology. In the aftermath of the Fukushima Daiichi disaster, it is understood that public acceptance of nuclear technology plays a central role in the decision-making process regarding systems operations and capacity deployment policies, even outside of the country where the incident occurred. For example, Germany decided to close half of its plants after the catastrophic events of March 2011, and will close the remainder by 2022. Other countries, however, depend on nuclear technology, or a considering it as a viable alternative for sustainable power generation. Typical efforts on capacity deployment and siting of nuclear power systems in the literature do not account well for long-term (e.g., 40+ years) uncertain drivers. This work introduces a novel approach to nuclear power systems design and capacity deployment under uncertainty that exploits the idea of flexibility and managerial decision rules. Flexibility in engineering design—also referred as real option in design—is promoted as a means to deal pro-actively with uncertainty, and has been shown in many contexts to improve life cycle performance significantly as compared to standard design and systems evaluation methods. Decision rules can be described as "IF-THEN-ELSE" statements, and are captured in the model via non-anticipative constraints. New design and deployment strategies are developed and analyzed through a multi-stage stochastic programming framework based on sample average approximation. The proposed solution considers flexibility in terms of phased capacity deployment, in-site capacity expansion, and life extension, subject to demand and public

M.-A. Cardin (✉) · S. Zhang
National University of Singapore, 1 Engineering Drive 2, Singapore, Singapore
e-mail: macardin@nus.edu.sg

S. Zhang
e-mail: zhangsizhe@nus.edu.sg

W.J. Nuttall
The Open University, Walton Hall, Milton Keynes, UK
e-mail: William.nuttall@open.ac.uk

G. Auvray et al. (eds.), *Complex Systems Design & Management*,
DOI 10.1007/978-3-319-26109-6_31

acceptance uncertainty. The numerical analysis shows that the flexible design benefits from life extension flexibility most significantly. Flexible phased deployment and capacity expansion are also important when electricity demand is the main uncertainty driver considered.

System-Level Modeling and Simulation with Intel® CoFluent™ Studio

Anthony Barreteau

Abstract Intel® CoFluent™ Studio is a visual model-driven development (MDD) solution for creating executable specifications of complex systems. It can be used at any point of the project lifecycle for modeling and validating any electronic or information systems in any application domain: hardware block, software stack, System-on-Chip (SoC), mixed hardware/software embedded system, networked/distributed system, end-to-end Internet-of-Things (IoT) infrastructure and Big Data networks. Intel CoFluent Studio can predict performance data from the application and use cases model execution on a multicore/multiprocessor platform model. Intel CoFluent Studio is a system modeling and simulation toolset based on Eclipse. Models are captured in graphical diagrams using Intel CoFluent optimized domain-specific language (DSL) or standard UML notations—a combination of SysML and the MARTE profile. ANSI C or C++ is used as action language to capture data types and algorithms. Non-functional system requirements or model calibration data such as execution durations, power, or memory values, are added through model attributes. Models are translated into transaction-level modeling (TLM) SystemC code for execution. The SystemC code is instrumented and generates traces that can be monitored with various analysis tools. Fast host-based simulations allow designers to observe the real-time execution of their application models on multiprocessor/multicore platform models. Performance figures such as latencies, throughputs, buffer levels, resource loads, power consumption, memory footprint, and cost can be extracted.

We will present this system-level technologies and associated methodology with a poster. The scope of the poster is related to the two following topics in technical and scientific methods:

A. Barreteau (✉)
CoFluent Technology Center, 3 Rue Alfred Kastler, Immeuble 'Le Saphir', Nantes 44300, France
e-mail: anthony.barreteau@intel.com

G. Auvray et al. (eds.), *Complex Systems Design & Management*,
DOI 10.1007/978-3-319-26109-6_32

- Systems architecture (needs capture, requirements development, systems modelling, simulation, optimization, sizing and specification, architectural frameworks).
- Systemic tools (configuration management, system behaviour analysis tools, modeling and simulation tools, test management).

Keywords System-level modeling · Executable specifications · Use-cases modeling · Performance prediction

A Systemic Meta-Model for Socio-Environmental Systems

Jérôme Dantan, Yann Pollet and Salima Taibi

Abstract We propose a systemic meta-model for the sustainable simulation of socio-environmental complex systems. The approach presented integrates data uncertainty management, for both representing and manipulating rigorously quantities which may have a finite number of possible or probable values with their interdependencies. We also provide an operationalization of such models for both data retrieving, via an object-relational mapping, and model simulation, via series of triples, which are linked to examples in the field of agriculture.

J. Dantan (✉) · Y. Pollet
CEDRIC, CNAM, 292 Rue Saint-Martin, Paris 75003, France
e-mail: jdantan@esitpa.fr

Y. Pollet
e-mail: yann.pollet@cnam.fr

J. Dantan · S. Taibi
Agri'terr, Esitpa, 3 Rue Du Tronquet CS 40118, Mont-Saint-Aignan 76134, France
e-mail: staibi@esitpa.fr

G. Auvray et al. (eds.), *Complex Systems Design & Management*,
DOI 10.1007/978-3-319-26109-6_33

The Smart Door: An Example of System Engineering in Building Industry

Gauthier Fanmuy, Arnaud Durantin, Hugo Messicat and Bertrand Faure

Abstract Systems Engineering is now becoming mandatory to master complexity but also to develop innovative systems. Application of Systems Engineering requires the use of a methodology upon tool set. This paper is about the application of a Systems Engineering methodology from CESAMES on a small but complex system: an automatic sliding door in a building. We all experienced it: automatic doors have tendency to open inadvertently for example when pedestrian just walks by with no intention to enter the room. This is due to an old technological design: easiest way to decide to open the door is to detect a person in a trigger zone. With a system approach, the door could be nicely improved with great potential developments. This document explains how, and the method used to do it.

G. Fanmuy (✉) · A. Durantin · H. Messicat · B. Faure
Dassault Systèmes, 10 Rue Marcel Dassault, Vélizy Villacoublay 78140, France
e-mail: G4Y@3ds.com

A. Durantin
e-mail: ADN3@3ds.com

H. Messicat
e-mail: HMT1@3ds.com

B. Faure
e-mail: BFU@3ds.com

G. Auvray et al. (eds.), *Complex Systems Design & Management*,
DOI 10.1007/978-3-319-26109-6_34

Architecture Approach for Managing System Complexity Using System Dynamics

Wael Hafez

Abstract Complex systems are defined by their behavior such as being adaptive, non-linear, or emergent. According to System Dynamics, the behavior and capabilities of a complex system are based on the dynamics of the underlying system structures. The interaction (information exchanges) among the various underlying structures, the feedback among them and the information processing delays involved along those interactions determine thus the system behavior. Accordingly, changes in the system structure impact its complex behavior and changes in system behavior requires changes to the underlying structures. The current approach argues that capturing the dependency between structural changes and system behavior can enable a better system design and management. That is, managing the structural complexity of a system (managing the number of elements used, their variety and level of dependency) can enable a better management of the system complex behavior. Introducing an additional architecture view to the system design that captures system structural complexity enables the depiction of the behavioral-structural dependency and a better evaluation of different system designs and management approaches from a structural complexity perspective.

W. Hafez (✉)
WHA Research Inc, 309 Holland Ln 226, Alexandria, VA 22314, USA
e-mail: w.hafez@wha-research.com

G. Auvray et al. (eds.), *Complex Systems Design & Management*,
DOI 10.1007/978-3-319-26109-6_35

We Choose MBSE: What's Next?

Aurelijus Morkevicius, Lina Bisikirskiene and Nerijus Jankevicius

Abstract When the decision is made to choose MBSE or the task is given to investigate whether MBSE is worth the investment, a long journey begins. The journey that requires knowledge, patience, and guidance to make the paradigm shift (from document-centric to model-based SE) rewarding. The final destination of this journey is prove that MBSE is rewarding in the context of a particular organizational. There are many barriers on the way, such as rumours about unsuccessful applications, too little information available how to proceed, disbelief, and a cultural change. Nowadays, MBSE is enabled by Systems Modelling Language (SysML). However, SysML is neither an architecture framework nor a method. This opens discussions of how to start, how to structure the model, what views to build, which artefacts to deliver and in what sequence. This paper summarizes the experience of different MBSE adoption projects in a form of a new framework for MBSE. The framework is organized in a matrix view and intends to help MBSE pioneers to answer the question "what's next?"

A. Morkevicius (✉) · L. Bisikirskiene
Department of Information Systems, Kaunas University of Technology, Studentu str. 50, LT-51368 Kaunas, Lithuania
e-mail: aurelijus.morkevicius@ktu.lt; aurelijus.morkevicius@nomagic.com

L. Bisikirskiene
e-mail: lina.bisikirskiene@ktu.lt; lina.bisikirskiene@nomagic.com

A. Morkevicius · L. Bisikirskiene · N. Jankevicius
No Magic Europe, Savanoriu ave. 363, LT-51480 Kaunas, Lithuania
e-mail: nerijus.jankevicius@nomagic.com

G. Auvray et al. (eds.), *Complex Systems Design & Management*,
DOI 10.1007/978-3-319-26109-6_36

Towards Smart City Energy Analytics: Identification of Consumption Patterns Based on the Clustering of Daily Electric Consumption Curves

Fateh Nassim Melzi, Mohamed Haykel Zayani, Amira Benhamida, François Stephan, Allou Same and Latifa Oukhellou

Abstract This paper presents the application of clustering algorithms to daily energy consumption curves of buildings. Our aim is to identify a reduced set of consumption patterns for a tertiary building during one year. These patterns depend on the temperature throughout the year as well as the type of the day (working day, work-free day and school holidays). Two clustering approaches are used independently, namely the K-means algorithm and the Expectation-Maximization algorithm based on Gaussian Mixture Model (EM-GMM). The clustering results obtained with the two algorithms are analyzed and compared. This study represents the first step towards the development of a prediction model for energy consumption.

F.N. Melzi (✉) · M.H. Zayani · A. Benhamida · F. Stephan
IRT SystemX, 8, Avenue de la Vauve, 92400 Palaiseau, France
e-mail: nassim.melzi@irt-systemx.fr

M.H. Zayani
e-mail: mohamed.zayani@irt-systemx.fr

A. Benhamida
e-mail: amira.benhamida@irt-systemx.fr

F. Stephan
e-mail: francois.stephan@irt-systemx.fr

F.N. Melzi · A. Same · L. Oukhellou
IFSTTAR, 14-20 Boulevard Newton Cité Descartes Champs sur Marne, F-77447 Marne la Vallée, France
e-mail: allou.same@ifsttar.fr

L. Oukhellou
e-mail: latifa.oukhellou@ifsttar.fr

G. Auvray et al. (eds.), *Complex Systems Design & Management*,
DOI 10.1007/978-3-319-26109-6_37

Model Identity Card (MIC) for Simulation Models

Saïna Herssand, Eric Landel, Jean-Marc Gilles and Joe Matta

Abstract Modeling a complex system implies the integration of different simulation models in various fields of expertise. These models should communicate with each other to simulate the behavior of the whole system. In this multidisciplinary context, the actors involved in the modeling process should deal with three main problems. Firstly, in order to reduce ambiguity, they need a common vocabulary and format to describe their models in a less informal way. Secondly, in order to reduce the cost of lately correction, any potential incompleteness and inconsistency problems related to the models should be identified in the early phases of creation and integration of models. Thirdly, the characterization of simulation models should allow actors to reuse existing models more efficiently. In this poster, we propose a common framework called Model Identity Card (MIC) to specify and characterize simulation models contents and interfaces. This new concept is implemented in arKItect (a MBSE tool) to facilitate the knowledge sharing between different actors. It allows users to reduce time to get a correct model by checking the completeness and consistency of their models throughout the modeling process. An industrial test-study in automotive industry is presented to illustrate the interest of the proposed approach.

Supporting multidisciplinary vehicle modeling by Göknur Sirin supervised by Bernard Yannou —Châtenay-Malabry, École Centrale de Paris and Eric Landel, Renault.

S. Herssand (✉) · J. Matta
Knowledge Inside, 7B rue Jean Mermoz, 78600 Versailles, France
e-mail: saina.herssand@k-inside.com

J. Matta
e-mail: joe.matta@k-inside.com

E. Landel · J.-M. Gilles
Renault Group, 1 avenue du Golf, 78280 Guyancourt, France
e-mail: eric.landel@renault.com

J.-M. Gilles
e-mail: jean-marc.gilles@renault.com

G. Auvray et al. (eds.), *Complex Systems Design & Management*,
DOI 10.1007/978-3-319-26109-6_38

From City- to Health-Scapes: Multiscale Design for Population Health

Matteo Convertino

Abstract Reconciling the growing proportion of the global population that lives in urban centers with the goal of creating healthy cities for all poses one of the major public health challenges of the 21st century. Genetics has accounted for only 10 % of diseases, and the remainder appears to be from the interaction of multiple socio-environmental causes that potentially determine epigenetic changes leading to diseases. Therefore, quantifying the dynamics of socio-environmental factors and the environment-disease linkages is extremely important for understanding, preventing and managing multiple diseases simultaneously considering population and individual biological information of exposed and non-exposed individuals. This is particularly important for the aim of reprogramming health-trajectories of populations and developing/managing cities with a quantitative health-based design. Here we show how complex systems models, and specifically, dynamic network factor analysis (DNF), and global sensitivity and uncertainty analysis can map the exposome-genome-diseasome network (i.e., the macrointeractome), determine network factor metrics useful for urban design, and assess probability distribution of comorbidities conditional to exposure in space and time, respectively. These probabilities are useful to make syndemic predictions by for design of socio-technical and ecological systems and intervention strategies in existing cities. As a case study, we use the SHIELD study in Minneapolis focused on measuring children's exposures to multiple environmental stressors and related effects on respiratory health and learning outcomes. Results show the very high degree of directional interaction among exposure factors and their spatial heterogeneity coupled to bi-directionally interacting diseases. We find non-linear conditional probabilities of disease co-occurrence and context-dependent dose-response curves that manifest large health disparities in populations. We show that macro socio-environmental features are much more important than biomarkers in pre-

M. Convertino (✉)
Environmental Health Sciences and Public Health Informatics Program, School of Public Health, Biomedical Informatics and Computational Biology Program, Institute on the Environment, Institute for Engineering in Medicine, University of Minnesota Twin-Cities, Minneapolis, USA
e-mail: matteoc@umn.edu

G. Auvray et al. (eds.), *Complex Systems Design & Management*,
DOI 10.1007/978-3-319-26109-6_39

dicting disease patterns with a particular focus on respiratory diseases and learning outcomes. Urban texture results as the most important factors, thus, such metric should be clearly considered in the design of socio-environmental systems via a minimization of the systemic health risk.

The developed probabilistic models are extremely flexible for the analysis of big data, city health-scape predictions, and optimal management of communicable and non-communicable diseases in socio-ecological systems via systems design. The understanding of linkages between structural, architectural, social, and environmental factors at the population scale will allow designers, architects, engineers, and scientists to design communities—from the material to the city scale—in which population health is the central objective of the design process.

Erratum to: Designing Systems with Adaptability in Mind

Haifeng Zhu

Erratum to:
Chapter 'Designing Systems with Adaptability in Mind' in: G. Auvray et al. (eds.), *Complex Systems Design & Management*, DOI 10.1007/978-3-319-26109-6_20

Erratum DOI: 10.1007/978-3-319-26109-6_40

The spelling of the author name 'Haifeng Zlu' was incorrect. The name should read as Haifeng Zhu.

The updated original online version for this chapter can be found at DOI 10.1007/978-3-319-26109-6_20

H. Zhu (✉)
United Technologies Research Center, East Hartford, USA
e-mail: ZhuHF@utrc.utc.com

G. Auvray et al. (eds.), *Complex Systems Design & Management*,
DOI 10.1007/978-3-319-26109-6_40

MIX
Papier aus verantwortungsvollen Quellen
Paper from responsible sources
FSC® C105338

If you have any concerns about our products,
you can contact us on
ProductSafety@springernature.com

In case Publisher is established outside the EU,
the EU authorized representative is:
Springer Nature Customer Service Center GmbH
Europaplatz 3, 69115 Heidelberg, Germany

Printed by Libri Plureos GmbH
in Hamburg, Germany